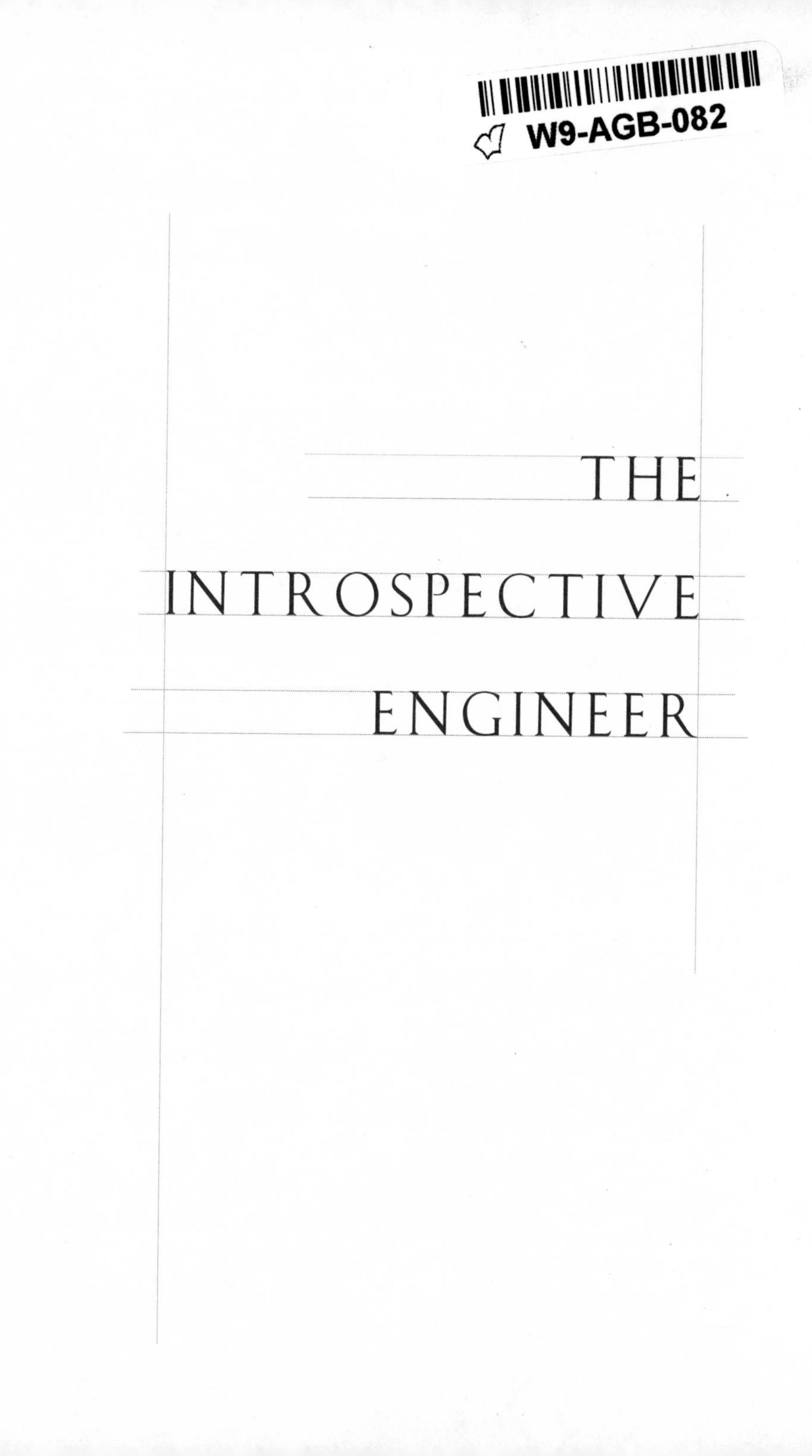
THE
INTROSPECTIVE
ENGINEER

ALSO BY SAMUEL C. FLORMAN

*Engineering and the Liberal Arts (1968)*

*The Existential Pleasures of Engineering (1976)*

*Second Edition (1994)*

*Blaming Technology (1981)*

*The Civilized Engineer (1987)*

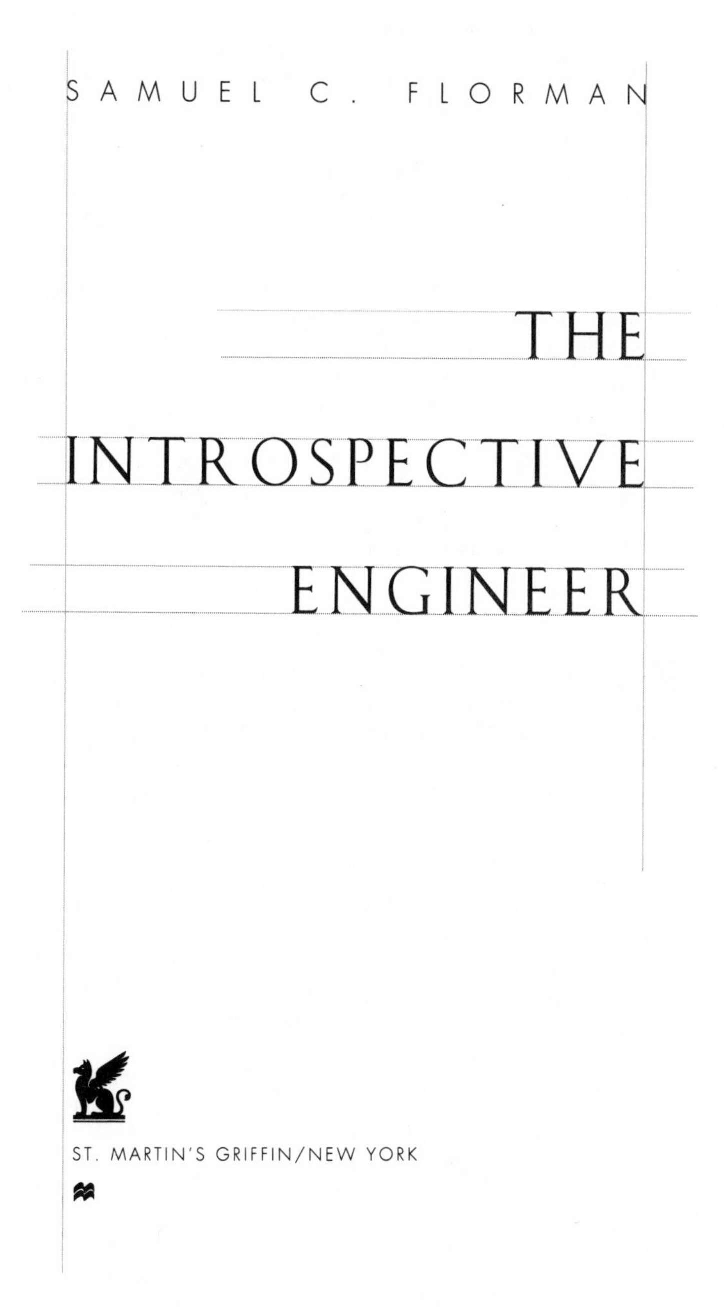

SAMUEL C. FLORMAN

# THE INTROSPECTIVE ENGINEER

ST. MARTIN'S GRIFFIN/NEW YORK

A THOMAS DUNNE BOOK.
An imprint of St. Martin's Press.

This book includes portions of columns that appeared in *Technology Review* from 1987 through 1995.

Part of Chapter 6 is adapted from "The Hardy Boys and the Microkids Make a Computer," *New York Times Book Review*, 24 August 1981.

Part of Chapter 10 is adapted from the George W. Woodruff School of Mechanical Engineering Annual Distinguished Lecture, delivered at Georgia Tech in 1991 and subsequently published in *The Bridge* (National Academy of Engineering), Fall 1991.



*Production Editor: David Stanford Burr*

*Design: Pei Loi Koay*

Library of Congress Cataloging-in-Publication Data

Florman, Samuel C.
The introspective engineer / by Samuel C. Florman.—1st ed.
p. cm.
"A Thomas Dunne book."
Includes bibliographical references.
ISBN 0-312-15152-7
1. Engineering—Philosophy. 2. Engineering—Psychological aspects. 3. Technology—Philosophy. I. Title.
TS157.F63 1996
620'.001—dc20 95-38836
CIP

First St. Martin's Griffin Edition: February 1997

10 9 8 7 6 5 4 3 2 1

*For Hannah and Lucy*

# CONTENTS

# PREFACE

Introspection is the act of looking inward, examining one's personal thoughts and feelings, or, more generally, "looking into or under the surface of things." Engineers—long said to be obsessed with materials and machines—are increasingly thinking in this mode. They are, however tentatively, seeking better understanding of themselves, their profession, and the role of technology in a rapidly changing world.

Broad, philosophical discussions are heard today at engineering seminars and meetings where they were rarely heard in the past. Professional journals print essays and letters to the editor expressing concerns about ethics, education, professionalism, and the relationship of technology to the general culture. Let me not overstate the case. The profession is not totally transfigured. Yet we have come a long way from the 1950s when an authoritative study of engineers found that "Constricted interests are apparent in their relative indifference to human relations, to psychology and the social sciences, to public affairs and social amelioration, to the fine arts and cultural subjects and even to those aspects of physical science which do not immediately relate to engineering."[1]

It could hardly be otherwise, considering the social turmoil of the 1960s, the environmental crisis of the 70s, the political upheavals of the 80s, and the evolution of a totally transformed global economy.

The introspection evolving among engineers parallels an increased interest in technology on the part of historians, ethicists, and social scientists. Indeed people in all walks of life are waking up to the importance of science and technology. I hope and expect that this will lead to a heightened interest in the profession of engineering.

In spite of widespread interest in the world of "high-tech," engineering itself lies in the backwaters of our national consciousness. This is regrettable. For reasons I will address in the Introduction—and throughout this book—it is important that engineering be more familiar, more discussed, and more deeply understood than it is in our society today.

This is the fourth book in a loosely connected series that began in 1976 with *The Existential Pleasures of Engineering*. In that work my main objective was to celebrate the intellectual and spiritual wonders of the technological impulse. In *Blaming Technology* (1981) I sought to defend engineers from the unwarranted allegations of their critics. In *The Civilized Engineer* (1987) I asked my fellow professionals, in effect, Why can't we be better than we are? In the following pages my approach is perhaps less focused, but every bit as heartfelt. I propose that we take the time to think about who we are, as citizens living in a "technological society," and—for engineers—as *the profession most essential to the well-being of that society*.

Since 1982 I have been writing columns for *Technology Review*, and some of those essays have found their way into this book, although extensively transformed and developed. This is true also for a few speeches that I have prepared for various occasions. However, a major portion of the book is completely new material.

We begin with introspection. But the implicit conviction of every engineer is that thought will lead to action.

# INTRODUCTION

*Where was the use, originally, in rushing this whole globe through in six days? It is likely that if more time had been taken in the first place, the world would have been made right, and this ceaseless improving and repairing would not be necessary now.*

—MARK TWAIN, *LIFE ON THE MISSISSIPPI*

If, to follow Mark Twain's thinking, the world had been "made right" in the first place, there would be little need for engineering. But because our planet is deficient in many ways—at least as a habitat for a rapidly growing population of more than five and a half billion humans—engineering is central to our lives.

One can deplore the current state of affairs and wish humanity could be what it was twenty thousand years ago, just a handful of folk traveling in small tribes and living by hunting, and gathering berries. (In fact, regrets about progress are to be found wherever progress has occurred. We find ample proof in the earliest myths, of which the one best known in the Western world tells of Adam and Eve, the Tree of Knowledge, and the expulsion from Eden.) But we cannot go back. We have no choice but to move ahead, coping as best we can with the difficulties that confront us: overcrowded cities, famine-plagued countrysides, pollution, disease . . . the all too familiar litany of human affliction.

Some people see our reliance on technology as cause for despair. Others—including many engineers—find inspiration in the challenge. Most of us are both inspired and despairing in varying degrees at various times. But like it or not, here we are, engaged in a struggle, not only for lives—which is difficult enough—but for lives that are worth living. We can hold our own in this struggle only if we develop ever more effective technologies. Even the problem of overpopulation requires us to make more technical progress: Experience shows that a modicum of prosperity is the necessary precondition for population control.

Indeed, experience shows that prosperity—not great wealth, but basic material comfort—is the necessary precondition for most of the things that human beings value. Somewhere along the Ganges, or high in the Himalayas, there may be sages who preach meek submission to the universe as we find it. But their followers appear to be growing fewer as the knowledge of ameliorative possibilities becomes widespread.

The United Nations Human Development Index rates nations on what they do to meet people's basic needs: keeping them healthy, raising educational standards, and helping them earn the income needed to make choices. Needless to say, the most technologically advanced nations achieve the highest ratings. While there is no automatic link between a country's per capita gross national product and its level of human development (for example, Costa Rica ranks higher in "human development" than Brazil, although the average income in Brazil is greater), essentially national wealth is the key to felicity.

One would like to include among "basic needs" the concept of political emancipation, and indeed at one time the U.N. Index did contain a "freedom" rating. This was dropped because of protests by some developing nations who feared that financial aid programs might be linked to political reforms. But even without a formal index, we know that the richest nations are also, by and large, the most democratic. Just about everything worth having begins with the development of material resources.

The term "technological fix" has become something of a pejorative in recent years, applied to thoughtless short-term solutions that result in long-term problems. Without pleading perfection for engineers—or for the people who often are in the position to tell engineers what to do—let me nevertheless suggest that at this moment in world history a few good technological fixes are just what we need. If, for example, we could provide ample supplies of pure water and wholesome food, conquer fearsome diseases, and develop new sources of energy—economical and environmentally benign—the elimination of mass misery could become a plausible objective.

"Happiness," to be sure, is a word that nobody can define to the satisfaction of all. But freedom from want, plus ample opportunity to "pursue" happiness—these are goals that are universally shared.

Compared to most of the world, the United States is a wealthy nation. But Americans, also, have problems that cry out for technical remedies: decaying infrastructure, environmental degradation, and so forth. And though we are rich, we are not nearly as rich, relative to other nations, as we used to be. If we could somehow manage to make better products—imaginatively conceived, expertly manufactured, and attractively priced—we would be well on our way out of the economic quagmire into which we have strayed.

For all our compassion and good will toward others, Americans are engaged in a fierce battle to maintain leadership in world commerce. We are committed to enhancing national "security" by means of industrial strength and—although the Cold War has ended—by means of superior armaments. Clearly we are ambivalent about technology. We think of it as a source of salvation for humankind; we see it as a means of keeping ahead of others in the pursuit of good things in a world of limited resources; we are apprehensive about its unanticipated side effects, social as well as physical.

It may be said that my emphasis on technology is simplistic. Others would argue the primacy of economics or politics. Do we need imaginative economic strategies? Absolutely. Politics is also important, political philosophy even more so. And it would help if human nature could be improved! But given the world as we know it, I suggest that prospects for national prosperity—and global salvation—rest heavily on our ability to do good engineering work.

Yet, engineering is not a word one is likely to hear in our communal discussions—for example, on Sunday morning television interview programs. Everybody agrees that we live in an era of "high tech," and that technology has changed our lives. Multimedia, virtual reality, information highway, genetic engineering—these are buzzwords of the day. But real engineers, the people who conceive of computers, and oversee their manufacture, the people who design and build information systems, cars, bridges, airplanes, and so many other things that are central to our lives, are nameless and obscure. Thirty-five percent of American adults have "no idea" how professional engineers spend their time, or think they run trains or manage boiler rooms.[1] Eighty percent of electrical engineers feel that the public has "no understanding" of what they do.[2] It is a very long time since Eiffel built his tower, since the Roeblings built the Brooklyn Bridge, and since Thomas Edison was said by the *New York Times* to be the most admired American.[3]

Does it really matter? So what if our popular heroes are athletes, rock stars, and business tycoons? So what if our television role models are lawyers and doctors, detectives and reporters, publishers and fashion models? Why should we pay more attention to engineers than we do to other essential but unsung people, like optometrists and bus drivers, or if you prefer a professional comparison, accountants and dentists?

One pragmatic answer is that because the quality of our lives depends in large measure upon the quality of our engineering, we must attract good people to the profession and keep them happy in it. At the moment we are not doing an adequate job of this. Enrollment in engineering schools grew steadily into the late 1980s, but has declined in the 90s, just when our need for technical ingenuity is greater than ever. Women and underrepresented minorities show profound indifference to the profession. This is particularly disturbing because, according to demographic studies, they will soon constitute a substantial majority of the workforce. We need a large and steady supply of engineers who are smart and well trained, and a goodly number who are ambitious and entrepreneurial as well. We need engineers who are idealistic and committed to public service. We need engineers who will become leaders, and we need potential leaders to decide that they ought to study engineering. If the profession is to grow, flourish, and be equal to the requirements of our society, engineering must be widely understood, appreciated, and esteemed. This can happen only if engineering is examined, discussed, and debated, if it becomes part of our everyday discourse, our art, and our popular culture. Our best young people need opportunities to be "turned on" to engineering. This is the first response to anyone who questions the importance of the issue.

Second, consider our politicians and decision makers. Most of them seem unclear about where engineering fits into the scheme of things—as well they might be with only one percent of the members of Congress, for example, having engineering backgrounds. We talk of economic growth, but our society is mainly in the hands of people who are, in the words of Labor Secretary Robert Reich, "pie-slicers" rather than "pie-enlargers." During the Cold War, much support for engineering came serendipitously out of budgets for national defense. Will support be there in the absence of direct military threat? Will the government invest in long-term technology development when short-term problems demand attention? And what of federal policies that indirectly affect technological advance? I refer to taxes, trade and investment strategies, antitrust restrictions, intellectual property rights, and product liability laws? Will politicians do what needs to be done if they—and their constituents—are uninformed about engineering, and confused about technology?

This brings to mind those people who worry about the adverse effects of technology. Isn't it important that they speak with some understanding rather than out of unthinking dread? Engineers have increasingly come to recognize the importance of politics in making technical decisions. They do not expect blind patronage; but they do hope for the rule of reason that is crucial to communal well-being. In a climate of confusion and anxiety, one fears that our representatives in government will fail to act constructively. Society needs informed politicians and enlightened citizens to support—and help guide—our technical enterprise. This is a second reason for us to be talking about engineering and technology.

There is a third reason that goes beyond pragmatism, beyond questions of global need and national strength. I suggest that our culture, to the extent that it ignores engineering—an essential element of its organic life—becomes impoverished.

When ice age peoples lived by hunting, they painted pictures of animals on their cave walls, a symbolic affirmation of the occupation that was central to their existence. In agricultural communities fertility rites evolved along with holidays to celebrate the harvest. When tribes went to war, they beat drums, danced, and made ferocious masks. Throughout history, societies have embodied their life-sustaining activities into art and ritual. In the United States we have a culture rich in art and music, literature and drama, both serious and popular. We have a wealth of newspapers, magazines, and cinema. Television is ubiquitous, bringing us serious documentaries and frivolous entertainment. In all of this, engineering is practically invisible. Yet engineering is a significant part of what we are, and an indispensable part of what we seek to be. Engineering can also be great fun—to learn about as well as to do. But, in spite of its great importance and inherent appeal, engineering is *taken for granted.*

Engineers themselves have not helped the cause, being as a group somewhat taciturn; although there are indications that this may be changing. Nevertheless, one hopes that engineers would not have to become orators or creative writers in order for their profession's importance to be recognized. (Cowboys were traditionally uncommunicative, yet they became central figures in American mythology!)

I do not look for romantic novels with engineer heroes—or heroines—nor a TV program called, *L.A. Engineer* as a few engineering colleagues have proposed. (Although, come to think of it, why not?) What I do hope for is a heightened awareness of engineering and its role in society, a modest objective, all things considered.

There are some hopeful signs. In the press, science writers are a relatively new presence, greatly enriching our cultural discourse. Some of them have drifted into technology and taken to discussing the work of engineers. (Science, by the way, has had its own image problems, and engineering's difficulties stem in part from a public that confuses engineering with science, or worse, sees it as subservient to its esteemed relative.) Financial writers, too, are increasingly interested in the work of engineers, recognizing the centrality of technology to the world of commerce. In academia, the history of technology has become a flourishing discipline, and related fields have arisen, such as STS (Science, Technology and Society), the Philosophy of Technology, and Engineering Ethics. On television, there have been a handful of "specials" with engineering content, and hopeful producers are planning others. As for creative literature, we've had Tracy Kidder's *The Soul of New Machine,* and a few other contributions that I discuss in Chapter 6.

A culturally enriching awareness of engineering and technology should also yield synergistic benefits for the first two goals argued above: the need for good young people to enter the engineering profession; and the practical need for leaders, as well as the public in general, to become knowledgeable about technological issues.

There is yet one more argument to support thinking about—looking into—engineering. I believe that the engineering cast of mind, a particular way of approaching problems—in short, the engineering view—has much to contribute to our society. Our public debates are too often characterized by passion, ill will, and distortion of the facts, sometimes intentional and sometimes unwitting. Engineers are trained to solve problems, adhering to facts and the truths of experience, shunning personal sentiment, or at least recognizing it for what it is. Engineers do not expect to find perfect solutions, because in their work there usually are none; they seek optimum solutions, given constraints of time, materials, and money. Their objective is to get a product "out the door," on schedule and within budget. They have to take human nature into account, considering what happens at a machine's "user interface." They are better than they used to be at predicting the environmental effects of what they do. In addition to being problem-solvers they are also—the best of them—imaginative creators, inventors, discoverers of new paths. They are realistic but not defeatist. If we are to prevail over the difficulties that beset us, this approach to problems, this engineering view, must percolate into the perspective of every citizen, and particularly into the outlook of our leaders.

A generation ago the engineering view could be summed up in the famed Seabee motto: "Can Do! The difficult we do immediately. The impossible takes a little longer." Engineers, like everybody else nowadays, have been sobered by the daunting realities of exploding populations, limited resources, and an environment that is more fragile than many had supposed. But the Can Do spirit is still very much alive. Within the parameters of the possible, engineers are willing to take on all problems, and to join in the struggle to improve the world, which as Mark Twain points out, doesn't seem to have been "made right."

A philosopher might speculate that the world was made in just such a way as to present us with challenges, to make us become engineers. If this is the case—if it is human destiny to become technological—then engineers have all the more reason to feel fortunate.

Technology and technologists, engineering and engineers—the terms and the topics mix and merge, with ripples of influence radiating and intersecting. I have divided this book into two sections, one dealing with engineering as a force in the world, the other with engineers; but there is much overlap.

Enough, then, of introduction. I began with a quote from Mark Twain's *Life on the Mississippi*. Let's proceed in the spirit of Twain's no-nonsense technologist hero, the Connecticut Yankee in King Arthur's Court: "To business now, and sharp's the word."

# I

# ENGINEERING

CHAPTER 1

# THE END OF COMPLACENCY

## SHOCK AND AWAKENING

Late in the summer of 1987 I accepted a book review assignment from the editors of a technical journal. The object of my critique was called *Strengthening U.S. Engineering Through International Cooperation: Some Recommendations for Action.* Although the title was long, I had been assured that the text was short—merely sixty-eight pages—and that the review was to be no more than five hundred words. This didn't sound like very hard work, and the topic seemed vaguely interesting, so I looked forward to the task with equanimity.

When the booklet arrived, however, and I glanced through it, my heart sank. It was a report by the Committee on International Cooperation in Engineering, an eminent group established by the National Academy of Engineering and the Office of International Affairs of the National Research Council, and at first look the report appeared to be—how can I put it?—drab, lackluster, let's just say dull. Nevertheless, a commitment is a commitment, so I had to follow through.

I set to work in mid-September, and I recall that at the time one event was totally dominating the news, claiming magazine covers and daily headlines. This was the debate about whether Robert Bork should serve on the U.S. Supreme Court. The Senate committee hearings on the Bork nomination lasted for two weeks, and wherever one went during that time, passionate arguments could be heard about politics and constitutional law. The only concern that ran those hearings a fairly close second—at least in the circles I frequent—was whether or not the New York Mets were going to overtake the St. Louis Cardinals in the race for the National League pennant.

As I started to work on my review of the NAE committee report, and as I started to weigh the significance of what I was reading, I felt that the ground was shifting beneath my feet, that tremendous upheavals were taking place, and that nobody was paying any attention. This calm, deliberate, dull booklet contained shocking news and dire warnings.

As the report made clear with a few key statistics, the United States was no longer the undisputed world leader in the realm of engineering research. We were no longer the brightest, most inventive, most creative people in the world. Of course, by 1987 I had heard a lot about foreign industrial competition. Everybody knew about Japanese cars and television sets. But I had assumed that at least in research the U.S. was still the leader by far. I thought, in fact, that engineering research was supposed to be our ultimate salvation. According to conventional wisdom, other nations may be good at making things, but we're better at devising new ways to make things and also at dreaming up new things to make.

Well, apparently that was no longer the case. It appeared that in at least 34 important disciplines comparable or superior technology was being created abroad. These included artificial intelligence, robotics, systems engineering and control, optoelectronics, combustion and engine technology, high speed rail, and nuclear plant safety. According to the findings of the NAE committee, the U.S. was not only losing its competitive advantage, the nation's leaders were not even keeping informed about what was happening elsewhere in the world.

The report suggested that in addition to quickening the pace of our own research activities, we needed to shed our self-satisfied complacency. It recommended that the National Science Foundation and other concerned agencies support American engineers in fellowships and sabbaticals abroad, sponsor domestic lecture tours by distinguished foreign experts, gather technical information worldwide, and encourage increased participation in international standards development. It further suggested that engineering students be given more opportunities to learn foreign languages and to study in other nations. And further still, it urged U.S. corporations and professional societies to become more aware of the need for international activities.

Yet everybody's attention was focused, at least for the moment, on the Bork hearings—on political intrigue and the niceties of constitutional law. The situation, as I perceived it, was positively unreal. I don't mean to imply that the Constitution is unimportant, or that freedom is not precious and worth debating passionately in a thousand forums. But, the essential precondition of individual freedom—the essential precondition of philosophy, free speech, and meaningful constitutions—is prosperity, or put most simply, material well-being. I have already argued this in the Introduction, and I will return to the theme again and again because it is so basic yet so often ignored. Plato could only hold forth in his academy because Athens was rich, and Athens was rich because Athenians were able technologists.

Don't people see, I mused, that our Constitution depends upon our technology? What would become of the hallowed American system of justice if our trade deficit continued to rise, if our economy became depressed, if we suffered from extensive unemployment, if our poor lost all hope and our workers became desperate? What would become of our fine judicial theories if the middle classes fell prey to uncertainty, and if the dispossessed began to riot? Very quickly we would begin to hear some new and not very palatable interpretations of our constitutional rights. Wherever we look in the world we see that freedom cannot exist independently of technological progress. Technology does not in itself create freedom; but freedom cannot exist without material comfort, and material comfort in today's world depends upon technology.

Above the portals of the New York State Supreme Court building in lower Manhattan there is an inscription that states: "The true administration of justice is the firmest pillar of good government."

Very nicely put. But as every engineer knows, pillars are worthless unless they rest on firm foundations. Technology is the foundation on which rest the pillars of justice and good government.

Thus ran my thoughts as I tried to fathom the meaning of that sixty-eight-page pamphlet. For a while I took to carrying it around as a sort of Bible, and even tried quoting from it to a few of my friends. But for the most part they only called me Chicken Little and went back to talking about the Supreme Court—and about the final exciting days of the baseball season.

October arrived and three things happened: I completed my review more or less on schedule[1], the Bork nomination failed in the Senate, and the Mets lost the pennant to the Cardinals. Well, life goes on, and I started to think about other things.

But almost immediately there arrived in the mail an invitation to participate in a colloquium sponsored by the American Association for the Advancement of Science. Each spring this organization brings together a number of specialists from government, industry, and academe to discuss the forthcoming federal budget for research and development. Discussions about the federal budget are not high on my personal list of priorities, and my initial reaction was, good grief, what could be more tedious? But this reminded me of the way I had felt when I first saw that dull report from the National Academy of Engineering. I looked at the invitation again. Budgets may appear to be boring, but they are important, just like engineering research is very important, and in fact the two topics are intimately related. So, in the spring of 1988, I went to Washington to hear discussions about the federal R&D budget.

The event, for all its importance, cannot be described as scintillating. As I sat there among congressional staff members and representatives of executive agencies, and watched columns of numbers projected on screens in darkened rooms, more than once my eyes started to glaze over. And later when I returned home and repeated some of the things I had learned about the R&D budget, I was greeted with total indifference. Again I was made aware of how little the average person knows—or cares—about how engineering fits into our national life.

Yet, the AAAS colloquium helped to alleviate the shock I had felt the previous summer in my encounter with the NAE report on engineering research. The United States may be facing fierce international competition, but it is far from a complacent and helpless giant. I was relieved to discover that there are people in Washington—and elsewhere—who recognize how crucial science and technology are to national vitality, and who are helping to allocate resources accordingly.

## IMPORTANT LETTERS: R&D

Research and development—the familiar R&D—is only one aspect of engineering, engaging less than 40 percent of American engineers (See Chapter 7). But it is the technological frontier, the place where we grapple with the unknown, and seek to devise the artifacts of the future.

Formal studies support the hypothesis—which in a way is intuitively obvious—that R&D is essential to the well-being of our society. Industrial innovation is central to wealth creation and economic growth, and R&D is a critical element of industrial innovation.[2] During the past fifty years "technological progress" has been responsible for about 40 percent of the productivity gain in the United States.[3]

Yet the average informed citizen knows hardly anything about what R&D is. During my visit to the AAAS colloquium, I realized that I myself had only the most superficial understanding of this vitally important topic.

I've met dozens of science and engineering professors, and graduate students, "doing" R&D on numerous campuses. I know that they get "grants," usually from some federal government agency, although occasionally from industry or nonprofit foundations. Their projects are sometimes highly theoretical ("Two-Phase Potential Flow" or "Modeling of Microstructural Evolution in Thin Films") and sometimes emphatically practical ("Microwave-

Induced Hyperthermia for Deep-Seated Tumors" or "Geophysical Surveys of Toxic Waste Spills in Interior Alaska"). I've met scientists and engineers who work at government laboratories, designing atom bombs, photovoltaic cells, or glucose monitors; also those who work for industry, at the fabled Bell Labs and elsewhere. I know that most large manufacturing corporations have R&D activities, for how else do they "invent" their products and learn to put them together?

But is this crucial activity appropriately supported? Who does it, who pays for it, and most important, where does it stand in our allocation of national resources?

Consider the year of this writing, halfway through the final decade of the twentieth century. The general profile, in constant dollars, has not changed substantially for a number of years.[4]

In fiscal 1995 the Gross Domestic Product (GDP) of the United States was just over $7.0 trillion. The federal government spent $1.54 trillion while earning $1.35 trillion in receipts, showing the deficit pattern with which we are familiar. The total committed to R&D in the U.S. was $188 billion. The main sources of this money were:

| | |
|---|---|
| The Federal Government | $73 billion |
| Industry | $107 billion |
| Universities, Colleges and other nonprofits | $8 billion |
| TOTAL | $188 billion |

Industry, in addition to its own commercial work, performs much of the government-funded R&D, as do the universities and colleges. Also, academe performs work for industry. But the key concern here is the source of the money.

The federal expenditure, when analyzed by agency funding, breaks down as follows:

| | |
|---|---|
| Defense Department | 49.9 percent |
| Health and Human Services | 16.1 percent |
| NASA | 13.0 percent |

| | |
|---|---|
| Energy | 9.1 percent |
| National Science Foundation | 3.4 percent |
| Agriculture | 2.1 percent |
| Commerce | 1.7 percent |
| Transportation, Interior, EPA, etc. | 4.7 percent |
| TOTAL | 100.0 percent |

The total allocation for defense, however, is not defined by the Defense Department allocation. If one counts military portions of the NASA and Energy budgets, the figure becomes 53.5 percent. (This is less than during the Cold War, when about two-thirds of the federal R&D budget was allotted to defense.)

At first look, I am pleasantly surprised to find that the federal R&D expenditure of $73 billion represents almost 5 percent of the $1.54 trillion federal budget. This seems reassuringly judicious until one considers that if 53.5 percent of the federal funding is dedicated to defense, then government expenditure on *civilian* R&D is only 2.2 percent of the federal budget. And, assuming that industry-funded R&D is 90 percent nonmilitary[5], *the total national investment for civilian R&D* (about $138 billion) is seen to be slightly under 2 percent of the GDP.

Is this a lot or a little? One unsettling indication is to compare the U.S. with Japan and Germany where the figure stands close to 3 percent. In absolute dollars, we spend more than any other nation. Yet our competitors spend proportionately more.

Of course, military R&D is not without value to the civilian sector. Claims may sometimes be overstated, but the benefits do exist. Not only do products developed for weapons turn out to have everyday application, but, more importantly, the military has long supported "pure" research, with general benefits accruing to society. This is particularly true in the realm of computers. When in the 1960s Douglas C. Engelbart, a computer scientist at the Stanford Research Institute, established the Augmentation Research Center, dedicated to developing human-computer interfacing, his work was funded by the U.S. Department of Defense. And when, a decade or so later, the counterculture-type engineers at Apple Computer Co. developed the "user-friendly" Macintosh, they relied heavily upon the techniques thus created with military support.

The world-renowned American R&D establishment is in large measure a product of the Cold War. With the ending of the Soviet threat, there was fear that support for the enterprise would diminish. Yet both the Bush and Clinton administrations set the objective of maintaining federal commitment to R&D, while reducing the military component to 50 percent. Another goal announced by leaders of both parties was gradually to increase the total spent on civilian R&D from 2 percent of the GDP to 3 percent, comparable to our leading commercial competitors.

However, the 1994 Republican victory in Congress called into question all previous strategies. The budget plan passed by the House of Representatives in May 1995 would have cut the civilian portion of federal R&D by 35 percent over a period of five years (after accounting for inflation), rousing the *New York Times* to editorialize:

> The party that preaches cost-benefit analysis for Federal agencies ought to practice what it preaches. . . . Knocking out innovative research can lead to stagnant productivity and growth. By that calculation, the House plan is an irresponsible gamble.[6]

However, shortly after the budget bill was voted, House Speaker Newt Gingrich, according to *Science* magazine, "privately delivered a surprising message to a handful of key legislators: Don't pull the purse strings too tight on federal research programs."[7] Happily, after the Senate had its say and negotiations with the White House were concluded, the final budget cuts were not nearly as severe as had been feared.

Nevertheless, in the tight-budget atmosphere that will likely prevail for the foreseeable future, there will be further battles in the R&D arena. Even people who would defend and enhance the federal R&D program are often at odds with each other about priorities.



One dispute pits champions of "big science" against supporters of "small," with the most significant recent loser being the gigantic Superconducting Super Collider, cancelled in 1994.

Perhaps the fiercest debates entail questions of "pure" R&D versus "practical" or "strategic." "Pure" or "basic" scientific research relies heavily upon government support. This time-honored concept is being questioned by people concerned about the crisis in industrial competitiveness. They suggest that government investment in R&D should yield not merely long-term, trickle-down returns, but more immediate returns by way of saleable products and increased employment, particularly in manufacturing.

A leading proponent of the new approach has been Senator Barbara Mikulski (D-MD), formerly chair of the Senate Appropriations Subcommittee that oversees Independent Agencies, including EPA, NASA, and the National Science Foundation (NSF). In calling upon the NSF to allocate 60 percent of its resources to funding research in "strategic" areas (such as biotechnology, infrastructure, advanced manufacturing, and high-performance computing) Ms. Mikulski said: "We're looking for ways to satisfy the immediate, compelling human needs of our society, while also planning for the long-range needs of the nation."[8]

As one might have guessed, opposition to Ms. Mikulski's "new paradigm" came first from scientists who do pure research (and who compete for funds with engineers who do more applied research and development). A more virulent protest came from people who feel that government should not intrude on industry's turf. The Mobil Company, for example, in newspaper advertisements run across the country, warned that "Congress should not be cooking up an increase in spending for the purposes of advancing whatever technology may catch its fancy." The business community's fear of a governmental "industrial policy" or "technology policy" has a long history. Some hard-liners reject any planning at all as an attack on the cherished system of allegedly free enterprise.

It is easy to see that when the federal government embarks on industrial ventures it can make mistakes (notably the breeder reactor and synthetic fuel production). On the other hand, government support for aeronautics, semiconductors, computers, and satellite communications has paid off handsomely for the national economy. It seems short-sighted to reject out of hand all federal support for commercial technologies just because of difficulties inherent in the process. It seems particularly inappropriate to disparage new government approaches at the very time that private industrial research is stagnating. From 1980 to 1985, spending on American corporate research grew at an annual rate of 7.5 percent in constant dollars. Since then, growth has barely kept ahead of inflation, and this at a time when foreign competitors have been increasing their efforts.

All indications are that "national industrial policy" will be central to our future debates. Already, as Robert M. White, past president of the National Academy of Engineering, has observed: "Just about every think tank, government agency, congressional committee, professional society, university, and industry association has issued a white paper, a report, a book, or a manifesto on the issues."[9] The Clinton administration pledged to work for technological advancement, with the government playing a prudent but significant investment role. Republicans see the world somewhat differently, stressing "incentives" to industry. One must hope that debates will focus on means rather than ends, and that the nation's commitment to R&D will not be thwarted. Actually, we must do more than hope: We must persevere in supporting the cause we know to be important to national well-being.

In all this discussion of budgets, agencies, corporations, and policies, we should not forget that engineering is the topic, and that real technologies are the issue. As a brief interlude, and in an effort to evoke a sense of reality, let's recall just a few of the basic technologies on which engineers are working: robotics, smart roads, biotechnology, machine tools, magnetic-levitation trains, fiber-optic communications, national computer networks, batteries, computer chips, sensors, computer aided manufacturing, advanced composite materials, artificial intelligence, digital imaging, and data storage. Let's consider some of the traditional industries in which R&D is ongoing: aerospace, automotive, chemicals, electrical products, food, fuel, health care, housing, leisure time products, manufacturing, metals and mining, office equipment, packaging, paper, semiconductors, and telecommunications.

My concern, at the moment, is not with any particular technology, nor with any particular technology policy or program—although there are many that pique my interest. Most important is learning—as I did at the AAAS symposium in 1988, and since—that despite public obliviousness, many smart and dedicated people are involved in supporting the cause of technology.

It is difficult to assess, literally from month to month, the health of the U.S. R&D enterprise. Was the 1987 NAE report overly pessimistic? Since my quasi-apocalyptic experience with that document, the general outlook seems somewhat improved. I pick up the Sunday newspaper and read that the U.S. economy is "Back in the Driver's Seat." "After more than a decade of painful change and dislocation, many American industries are leaner and nimbler, and others have seized the leadership of the sophisticated technologies that are ushering in the information age."[10] According to the 1995 Biennial National Critical Technologies Report, "the United States is well positioned in those technologies which are deemed to be critical to the nation's economic prosperity or national security." However, "the size of the U.S. lead has either declined or remained constant."[11]

Even if we are indeed holding our own in competition with other industrial powers—and tomorrow's reports may suddenly tell a different story—we certainly aren't holding our own in coping with the needs of many of our citizens, to say nothing of others throughout the world.

Nor can we be satisfied with the electronic marvels of an "information age," dazzling as they may be. Alvin Toffler may hail the coming of a "Third Wave," claiming it is equivalent to earlier revolutions in agriculture and industrial machinery.[12] But unless silicon chips and fiber-optic cables can be made to feed, clothe, warm, and house the needy, Mr. Toffler and his followers would be wise to mute self-congratulatory celebrations.

The ultimate technological revolution—which will feature affordable sources of energy, new materials, ample food and water, and a wholesome environment—remains a faraway dream. It will require legions of insurgents—led by creative and committed engineers—to make it a reality.

## THE END OF HISTORY?

My sojourn into the world of Research and Development leaves me—after some shocks and scares—guardedly optimistic. The public may be largely out of touch with vital technology issues—more concerned about politics, baseball, and "democracy"—yet somehow far-sighted support for engineering has been forthcoming. However, in the absence of military threat will such support be maintained and revitalized? For every moment of encouragement there seems to be a countervailing occasion for misgiving.

For example: One would have thought that the collapse of communism—an event clearly linked with the inability of centrally planned economies to manage technology successfully—would have brought about a new respect for engineering, particularly among statesmen, intellectuals, and other opinion makers. But the winning of the Cold War, and the events that ensued, have provided new opportunities for scholars to turn their back on the technological enterprise and to concentrate on ideology.

This was illustrated with particular force by the intellectual ferment surrounding "The End of History?", an essay that appeared in the Summer 1989 issue of *The National Interest.* Written by Francis Fukuyama, a State Department official, this article quickly became, in the words of a *New York Times* editor, "the hottest topic around."[13] Although the piece was debated at length in the press and in intellectual journals, commentators concentrated on Mr. Fukuyama's political and philosophical statements, remaining strangely silent about his attitudes toward technology. I found those attitudes nothing less than appalling, particularly considering that the author held a responsible position in government.

The idea implicit in the article's provocative title is that because Marxism-Leninism is disappearing "as a living ideology of world historical significance," history as we have known it is coming to an end. Western democracy has prevailed; the once all-absorbing conflict is concluded. Although the triumph over totalitarianism is cause for celebration, Mr. Fukuyama was at the same time moved to melancholy. In a passage that was widely quoted, he observed: "The end of history will be a very sad time. The struggle for recognition, the willingness to risk one's life for a purely abstract goal, the worldwide ideological struggle that called forth daring, courage, imagination and idealism, will be replaced by economic calculation, the endless solving of technical problems, environmental concerns and the satisfaction of sophisticated consumer demands." Further, there looms before us the "prospect of centuries of boredom."

Critics of this essay concentrated on its quotient of fantasy. How can one say that democracy has prevailed once and for all? Won't ideological disputes continue to flare wherever people try to govern themselves? The events of ensuing years—in Bosnia, Rwanda, and elsewhere—showed that such doubts were clearly valid.

What concerned me, however, was the author's gratuitous disparagement of engineering, and the fact that none of his critics picked up on it. In equating "the endless solving of technical problems" with "centuries of boredom" Fukuyama voices the intellectuals' contempt for technology, a superficial and arrogant view of life that is at least as old as classical Greece.

If the threat of boredom were ever to hang heavy over humanity (a far-fetched conceit, but let's assume it for the sake of discussion), science and engineering—along with the arts—would provide the antidote. Should it not be obvious to all thinking people that solving technical problems is a quintessentially human thing to do, is in fact one way of defining what human beings are? Even if all our social and political problems were to be resolved, science and engineering, by their very nature, remain always on the edge of new frontiers. The universe presents itself to us eternally as a mystery to be studied and acted upon. Albert Einstein wrote of the "inner freedom and security" he found in scientific contemplation. Rudolf Diesel said that the conceptual breakthrough that led to the invention of his engine filled him "with unutterable joy." When it comes to ennui, Mr. Fukuyama confused the remedy with the affliction.

My argument, however, goes beyond the philosophical question of what constitutes meaningful and gratifying activity. If we nonchalantly dismiss technology as a minor factor in the scheme of things, we risk re-energizing those very forces of evil that with the fall of communism have somewhat abated. Once the euphoria of freedom has run its course in Eastern Europe, and eventually in China and elsewhere, the problem of how to order our lives, politically and socially, will quickly return to the fore. And here technology is not merely a peripheral factor to be considered casually by civil servant diplomats. Technology lies at the heart of the matter. Freedom by itself is not what counts. People crave freedom of expression, but not if it comes with the freedom to starve, or even the freedom to struggle for subsistence while a few of their neighbors enjoy luxuries.

Communism failed primarily because it did not provide a nourishing climate for technical progress. The East Germans crossing the Berlin Wall spoke less about freedom of speech than they did of the cornucopia of goods they saw displayed in the West. Asked about the new era of emancipation, a Bulgarian observed that "free opinions, or for that matter elections, do not put bread on the table. The important thing is not who wins an election but that I have enough money to buy things, and enough things to buy."[14] If the newly established governments fail to fulfill the hopes raised by Western technology, democracy will once again find itself under attack from communist forces, or even worse, from fascist demagogues.

How are science, engineering, and industry to be supported and improved, and their fruits equitably distributed? What of technology transfer, technical education, environmental concerns, and turning swords into plowshares? These are the questions that need to be addressed in the post-communist era. Granted, ideology is also important. But divorced from technology, or even given priority over technology, well-intentioned political sentiments will be of no avail.

Mr. Fukuyama, as part of his official duties, prepared "talking points" for the secretary of state's meetings with foreign leaders. It is frightening to think that his condescending ideas about "technical problems" might represent attitudes in high places. Unless the world's leaders recognize that a strong technology is a necessary (although not sufficient) precondition of freedom, we may find ourselves, not bored by the end of history, but dismayed at having to relive it.

We have reached, not the end of history, but the end of complacency. The times require constructive initiatives by government and industry, plus creative, ingenious technical work by those trained to perform it, especially our engineers. Good things will not just happen spontaneously. There must be public support and private motivation. Underlying all, creating energy and incentive, there must be a widespread affirmative feeling about the engineering enterprise. There are reasons to hope that this will emerge as more people come to recognize our true societal needs.

Yet, there are many uncertainties in the public mind about engineering and technology, and it would be counterproductive for engineers to ignore or belittle them. It would also be foolish, because most of these uncertainties are founded in legitimate concerns.

CHAPTER 2

# ALLAYING APPREHENSIONS

## FALSE PROPHECY OF A STREAMLINED FUTURE

What will the world be like after it has been saved by engineers? During most of the twentieth century, technological progress has been identified in the public mind with streamlining and slick design. The term *modernism* has a certain appeal, but it also fills the thoughtful person with foreboding. When we think of modern furniture, modern architecture, modern art—modern design in general—there is an association with mathematical austerity. We sense the increase of smooth artificiality and the loss of natural luxuriance. Is there indeed a technological imperative that allows us to prosper only at the cost of traditional beauty and grace? Shall we preserve our lives only to find we have lost our souls?

This question occurred to me with a particular intensity in the Spring of 1994 when the TV series, *Star Trek: The Next Generation,* was cancelled at the end of its seventh season. The cancellation was attributed, paradoxically, to the show's great popularity. It was ripe for repackaging and sale. Local TV stations bought its 182 episodes, planning to run them every night of the week for months on end, and then start the cycle over again. Six months after the airing of the last episode (which attracted 31 million viewers) a movie based on the TV show (the seventh in the Star Trek film series) was released. Two new spin-off programs were soon on the air. Three decades after the original *Star Trek* series

(NBC 1966–69), this incredibly successful enterprise had become, according to a TV editor, "that rarest of show-biz flowers, a franchise, something Hollywood studios take great care to nurture and cultivate."[1]

Of all artistic visions of the future, *Star Trek* seems to have had the greatest public appeal. Yet what do its creators have to tell us about the world of tomorrow? In the original 1960s episodes—which established the essential ambience and themes—they show us a future lived mostly in outer space, where the people, as if to match the forbidding environment, are marked by a cool aloofness. There are some exciting battles, to be sure, and lots of fascinating gadgetry, but nobody seems to have much fun. There is proclaimed love and occasional sex, but no passion; there are jokes but no geniality. The good guys mouth noble sentiments, but they do not seem to savor life. The people with the most zest are the villains, and indeed if life is as bland as it appears to be, an individual with spirit might have good reason to become a rebel. This holds true not only for the early *Star Trek,* but also for *Star Wars, Dune,* and other creations of the same ilk.

What do they *do,* these people of the future? Apparently they don't have to work. They don't seem to go on picnics, hikes, or bicycle trips; they don't dance or play basketball, read novels or give concerts. They fly their spaceships—which is good sport I suppose; but how much of that can one take? And their clothes! Caring neither for blue jeans nor elegant dress-up outfits, they seem to run around mostly in long underwear. This vision of the future is summed up in the look of a space ship command center: stainless steel and glass, blinking lights and whirring disks—bleak, stark, and barren. This is where we are headed, one might suppose, as engineers lead us into the next millennium.

Such a concept makes sense only if we extrapolate certain contemporary trends far into the future. But I believe that in espousing this vision our popular artistic creators have gone wildly astray. An environment of spaceports and domed metropolises is not the world toward which our technological society is headed. The slick, the clean, the "modern" may have its allure, but it fails to satisfy quintessential elements of the human spirit. And there is no reason to believe that the human spirit cannot stand up for itself, is not indeed already showing signs of rejecting the prophecies of science fiction.

The joggers who stampede past my door each morning speak to me in body language of a primordial zest for meadows and hills. Gardening is more popular than it ever was. Sales of canned foods have declined while the market for fresh produce has expanded. Carpentry and other crafts have had a renaissance. Clothing fashions are as frivolous as they ever were, with ruffles and bows coming and going and hemlines moving up and down from season to season. The environmental movement, born from the fear of health hazards, has expanded on the wings of an aesthetic revolution. People not only want unpolluted air and water, they want unpolluted landscapes. They want hedges and rivers and vistas. In their cities they want old neighborhoods restored. Parks, fountains, and brick-paved streets are in. Freeways and glass-walled canyons are out. The world of *Star Trek* is a fantasy not only for today but for all time. Such a world can never come into being because humans insist upon softness, spontaneity, and caprice. This is not just a matter of choice; it derives from what we are and how we have evolved.

Our stone age ancestors of twenty thousand and more years ago covered the walls of their caves with graceful paintings of animals. These ancient artists manifest the exhilaration of life in the forests and on the plains. Toolmakers, potters, and weavers of antiquity decorated their handicraft with symbols derived from the world around them: leaves, vines, insects, birds, and the like. As civilizations emerged, other shapes took their place in the human psyche: cultivated fields and arbors, houses and castles, town squares and markets, sailing ships and horse-drawn carriages. All were lovingly portrayed in tapestries and illuminated manuscripts, and all took their places in the decoration of everyday objects.

With the coming of machines and mass production, more new forms appeared: cylinders and piston rods, cogged wheels and rollers, suspension bridges and skyscrapers, eventually jet planes, rockets, and silicon chips. But these new shapes, which came very late in the evolutionary life of homo sapiens, did not suddenly change senses and tastes that had developed over millennia. The romantic, frivolous human heart was not transmuted abruptly just because the human brain conceived new products and processes or even because this amazing brain discovered certain shapes and relationships inherent in the structure of the universe. Tastes change, to be sure, but only within limits inherent in the species that we are. We have resisted efforts to create a geometrically defined environment, purified and stripped of natural flamboyance.

A major attraction at the Paris Exposition of 1867 was the locomotive *America*. Its cab was crafted of ash, maple, black walnut, mahogany, and cherry. Its boiler, smokestack, valve boxes, and cylinders were covered with a glistening silvery material. The tender was decorated with the arms of the Republic, a portrait of Ulysses S. Grant, and a number of elaborate scrolls. Other machinery of the day exhibited similar characteristics. Steam engines were built in "Greek revival" style, featuring fluted columns and decorated pedestals. On a printing press called *The Columbian* each pillar was a caduceus—the serpent-entwined staff of the universal messenger, Hermes—and atop the machine perched an eagle with extended wings, grasping in its talons Jove's thunderbolts, an olive branch of peace, and a cornucopia of plenty, all bronzed and gilt.[2]

It is little remembered today that well into the late nineteenth century most American machine manufacturers embellished their creations. While this practice pleased the public, some observers considered it anomalous. A writer in the British periodical *Engineering* found it "extremely difficult to understand how among a people so practical in most things, there is maintained a tolerance of the grotesque ornaments and gaudy colors, which as a rule rather than an exception distinguish American machines."[3] An exasperated critic for *Scientific American* asserted that "a highly colored and fancifully ornamented piece of machinery is good in the inverse ratio of the degree of color and ornament."[4]

By the beginning of the twentieth century, machine ornamentation yielded to clean lines, economy, and restriction to the essential. "Form follows function" became the precept of a new machine aesthetic. Creators of exotic contraptions like the locomotive *America* were accused of being sentimentalists, hypocrites and worse. Yet in their reluctance to give up adornment—ridiculous as it might have seemed—these designers were in fact expressing a discomfort we all share, an uneasiness in the face of mathematical severity.

The new machine aesthetic, the admiration of slickness and purity of line, spread from factories and power plants into every area of society. The term "industrial design" was first used in 1913, and by 1927 the famed Norman Bel Geddes was calling himself an "industrial designer."[5] During the twenties and thirties practically every human artifact was repatterned in the new mode. Lamps, tables, and chairs; toasters, refrigerators, and clocks; plates, goblets, and flatware—all were simplified, trimmed, and reshaped. Even the humble pencil sharpener did not escape; Raymond Loewy created a streamlined, chrome model in 1933.

Along with the revolution in style, came many theories about why it was happening—admiration and emulation of the machine being only one. The new simplicity, it was claimed, was democratic at heart, a rebellion against the baroque ornateness of older, autocratic societies. A more jaundiced view held that the new vogue was intended to distract the masses in hard times, or simply to help promote the sale of products by giving the machine a good name.

The increasingly popular style also evoked wry protest. In a *New Yorker* cartoon of 1932 the chairman of a corporate board of directors addressed his colleagues: "Gentlemen, I am convinced that our next new biscuit should be styled by Norman Bel Geddes."[6] Of course, biscuits were not recast in the modern mode; animal crackers retained their popularity, along with gingerbread men and Lorna Doones. For every streamlined, molded chair designed by such newly famous personalities as Charles Eames, consumers purchased dozens of overstuffed Victorian loveseats and Colonial rockers.

The struggle between modernity and traditionalism reached a peak of intensity in architecture. Out of the Bauhaus School in Weimar,

Germany, during the 1920s, there came a new movement, a "functionalism" that featured flat roofs, sheer facades, and exposed structural skeletons of steel and concrete. Ornamentation was scorned as extraneous and bourgeois. Walter Gropius, Mies van der Rohe, and Marcel Breuer were three of the stars of this new academy, and they were joined in philosophical union by the Swiss-born French architect, Le Corbusier, founder of Purism, who called the houses he designed "machines for living."

From its beginnings, the so-called International School encountered resistance among the very clientele it claimed to represent—the common people. A housing project for workers, sponsored by the Stuttgart government in 1927 and designed in the functional style by a team of modernist luminaries, did not appeal to its tenants. It was all very well for Gropius to maintain that these people were as yet "intellectually undeveloped," and for Le Corbusier to say that they would have to be "reeducated." The plain fact was that the residents preferred pitched roofs, eaves, cornices, lintels, capitals, and pediments. And once inside their barren boxes, they yearned for moldings, pilasters, mantels, and radiator covers (the coils had been left bare as honest, abstract sculptural objects). Nor were the residents comfortable with the uniformly white walls and the gray and black linoleum floors.

"Cool cubes" Tom Wolfe has called these earliest modernist dwelling spaces, and nobody has better described the distress that such architecture has brought even to the prosperous owners of costly, chic, custom-designed residences. "I once saw the owners of such a place," writes Wolfe, "driven to the edge of sensory deprivation by the whiteness & lightness & leanness & cleanness & bareness & spareness of it all."[7] Of course, while architectural magazines featured new "homes of the future" with flat roofs and large glass walls, most houses continued to be built in traditional shapes or in such comfortable new varieties as the split-level ranch. At the height of the modernistic vogue, refurbished barns still rated high on the scale of status symbols.

In commercial architecture, as opposed to residential, the dominance of functionalism was for a while almost uncontested. Shortly before World War II, Gropius, Breuer, and Mies van der Rohe immigrated to the United States, and in the building boom that followed the war their influence was everywhere. Under the aegis of such American converts as Philip Johnson, I. M. Pei, and Gordon Bunshaft, the typical office building became a glass box. This suited private developers who found that the boxes were economical to build, and it also pleased academicians and critics who took to quoting Mies's aphorism: "Less is more." But office buildings were just the beginning. Boxes of steel and glass and concrete began to appear everywhere, including such traditionalist bastions as the Gothic campus of Yale University.

Protest was slow in coming, but inevitably it grew. In 1954 Edward Durell Stone designed the American Embassy in New Delhi, a striking building sheathed in ornate terrazzo grills and set in a lavish water garden. In 1956 Eero Saarinen created a new TWA airport terminal in the shape of an eagle. These two mavericks were for a while scorned by the leading savants of the architectural world as were such hotel architects as Morris Lapidus and John Portman, whose facades, lobbies, and atriums featured flamboyant shapes, sumptuous textures, and luxurious plantings. But when Robert Venturi in 1966 wrote a book called *Complexity and Contradiction in Architecture* in which he called for "messy vitality" to replace modernism's "obvious unity," the counterrevolution had an intellectual theme to support its grassroots energy.

During the seventies, "postmodernism" became the rage, and architects once again graced their buildings with decorative columns, capitals, and cornices. The flat-topped office building slipped toward disrepute, and designers of commercial towers tried to outdo each other with domes, gables, pyramids, and spires. A San Francisco ordinance actually mandated that "all buildings should be designed to create a visually distinctive termination of the building facade." For engineers, the return to ornateness provided new design challenges. None of the engineers I know have ever insisted upon giving geometric imperatives priority over human tastes. Certain simple shapes have advantages of economy, but engineers do not insist upon economy when a client has other priorities. Engineers enjoy novelty and variety as much as other people, and perhaps more than most.

Resistance to "modern" design has been accompanied by opposition to new "artificial" materials. Plastics, first promoted as the basis of a utopian environment ("nothing to break, no sharp edges or corners to cut or graze, no crevices to harbor dirt or germs," crowed a 1941 article in *Science Digest*), quickly came to be the symbol of everything shallow, superficial, and phony. In 1969 the editor of *Modern Plastics* complained that although consumption of the product continued to expand, "plastics' reputation has remained about as low as it can get."[8]

Consumers still prefer leather shoes to Corfam, and leather-covered couches to vinyl. Cotton shirts, most people agree, are nicer than rayon. For high-quality clothing, pure wool is the fabric of choice; the polyester suit has become an object of derision.

Surely none of this has escaped the attention of utopia planners and creators of science-fiction movies. These savvy people see the appeal of traditional styles and materials. They assume, however, that technological progress will eventually overwhelm our nostalgic impulses. It is only, they seem to say, a matter of time.

In making such projections they are deluded on two counts. First, they underestimate human tenacity and the hold of the past on the human heart. Second, they ascribe to technology a quality of slickness and gloss that just isn't there. It is true that airplanes and rockets take on shapes that are expressions of aerodynamic formulas. But airplanes and rockets are only a small part—a very small part—of the world in which we live. Camping gear, to pick a mundane example, made of aluminum and nylon—two large energy-intensive technologies—helps make it possible for people to commune with nature in the most remote woodlands, to enrich their spirit, and refresh their psyche.

Not that forests are the ideal habitat for which the human soul yearns. I once heard Rene Dubos, a respected environmentalist, startle the audience at a symposium on ecology by speaking of the "pollution" caused by the regrowth of forests where small farms have been abandoned. People need vistas, he explained. Having evolved genetically in the savannah of East Africa, and socially in fields and towns, we are ill at ease when constantly hemmed in by trees and undergrowth—also in the barren expanses of limitless desert.[9] We are equally ill at ease, he might have added, when confronted with the mechanically slick settings of futuristic movies.

Several years ago I spent a day in Disney World. Invited to attend a conference, I arrived early and went immediately to the Magic Kingdom. I wandered through Adventureland, Frontierland, and Fantasyland, going from ride to ride. "Pirates of the Caribbean," "Tom Sawyer Island," "Snow White's Adventures"—I was as delighted as any child in the place. I finally arrived in Tomorrowland and with great expectation embarked on a ride called "Mission to Mars." It was a letdown, plain and simple. My young fellow passengers and I agreed that when it comes to fun and enrichment of the senses, space travel is overrated. At least it does not compare favorably to a trip over moonlit London with Peter Pan.

There will some day be manned spaceships voyaging to other planets, perhaps even—who can say?—to other galaxies; and I hope that one of them is named *Enterprise* in honor of the *Star Trek* craft. But I do not believe that our grandchildren, or their descendants for many generations, will live in the slick and streamlined environment that comes to mind when we think of rockets and space travel. The people simply will not have it so; and although engineering relies upon science and mathematical verities, in the end it responds to the demands of the human spirit.

## THE FEAR OF A HOMOGENIZED WORLD

Nevertheless, one can hear skeptics reply, there is small comfort in resisting the conquest of stainless steel and glass if all the while communities in every part of the globe are becoming more and more like each other. How are we to respond to the widespread concern that technology is homogenizing the world? As a disgruntled woman said to me at a cocktail party, "Bangkok is getting to look just like St. Louis!" Clearly, as more nations become technologically advanced, similarities begin to appear. Sameness is a price we pay for better shelter, sanitation, and transport, for the dissemination of refrigerators, radios, and washing machines. It is only natural to want health and comfort for all, but isn't it a shame—so goes the complaint—that this tends to reduce the differences that have distinguished peoples, and their habitats, from each other.

The reasons for the propagation of homogeneity are easy to identify. Efficient engineering solutions appeal to logical people everywhere, hence bridges, skyscrapers, and automobiles tend to resemble each other wherever they appear. Manufacturers in the most technologically advanced societies export their mass-produced merchandise to less developed lands, so tractors, locomotives, and computers travel in their original shapes and patterns, as tourists travel in their own clothes. Technical standards for tens of thousands of items—from nuts and bolts to mattresses—promote efficiency and economy, and so are adopted internationally. Cultural traits are spread by movies, magazines, and television, making Mickey Mouse and rock music common around the globe. Imitation plays a part in this dispersion, and less benignly, there is the pressure of salesmanship and advertising.

Having conceded these facts, it nevertheless appears that a powerful counterforce is at work, the effects of which have been too little noted. Our elegiac concern for the death of diversity may prove overstated. Modernism, like medical inoculation, contains within itself the seeds of antimodern sentiment. For whatever reason or reasons, people in many nations have taken to battling fiercely in defense of local customs. In Iceland the purity of the local language is protected by the government. In some Islamic lands the adoption of Western dress and customs is punishable by severe penalties. In every corner of every continent one finds renewed pride in traditional folkways, and the sponsorship by state and private groups of native dress, dance, architecture, and the like. Technology plays a positive role in such activities by providing wealth and leisure time. This makes for introspective consideration of alternatives. It also gives traditionalists the same educational and propaganda tools that contribute to change in the first place.

Such resistance, however, is only partly effective. It may prove as difficult to fend off the spread of blue jeans, for example, as it is to stop the proverbial rising of the tide. But the passion to preserve traditional customs is only part of the counterforce to which I refer. The advance of technology, *in itself,* creates new opportunities for diversity even as it appears to promote a bland sameness.

A characteristic of technologically advanced societies is the availability of cheap transport. Another, which stems from the division of labor, is a complex and fluctuating job market. A third is the opportunity for many individuals to accumulate capital, albeit for most people in modest amounts. What this means is that the average citizen can get a job, save a little bit of money, then travel to another place and get another job, or even embark on a new business venture. Such mobility is often decried by critics of technology who claim that it contributes to footlooseness and loss of roots, and further that the freedom provided is illusory, since wherever one goes the landscape is liable to be blemished by the same shopping centers, and the airwaves tainted by the same radio and TV shows. Yet there is a good deal of evidence that mobility contributes, not to sameness, but in a powerful and meaningful way to the very diversity whose purported loss is being lamented.

Consider some of the individuals involved in the rebirth of small food-producing ventures in New England: a former computer service manager from Manhattan who grows vegetables hydroponically in Lakeville, Connecticut; a schoolteacher from Ohio and a New York City tugboat captain who raise lamb on a farm in Putney, Vermont; a Canadian carpenter who grinds cornmeal in Adamsville, Rhode Island; an engineer from Brooklyn who quit his job in the defense industry to found a yogurt-producing farm in Wilton, New Hampshire; and dozens more.[10] Consider that while the number of large American farms has been declining, the number of small farms has increased. Exurban micro farming (annual income of up to $10,000) is a national phenomenon, much of it done part-time by white collar professionals, by recent immigrant families, and by retirees.[11]

Clearly there are many people who yearn for rusticity; and paradoxically it is a technologically advanced society that makes it possible for them to realize their aspirations. Once established in the country, they actively oppose real estate development, and form the core of those who would preserve the pastoral way of life.

Traveling in the opposite direction are the many thousands who flee what they consider to be the tedium of farms and small towns, seeking careers in industry and the arts, and reveling in the excitement to be found in large cities. Then there are the scientists gravitating to research centers, academics heading for universities, bureaucrats seeking centers of government service, and so forth. Beyond the attraction of specific vocations, there lies the allure of a more vague concept encapsulated in that recurrently heard term, "lifestyle." Mild climate attracts multitudes to the Sunbelt, while a craving for "laid-back" ways draws many to California. In extreme cases we find Pakistanis driving taxis (or studying engineering) in Cambridge (England or Massachusetts) while spiritually inclined Westerners seek enlightenment in the Himalayas.

Having arrived at the places of their choice, people show a remarkable determination to prevent those places from changing. Indeed, the combination of dedicated natives and idealistic newcomers synergistically amplifies the attitudes and mores that make each place unique. Further, as the global electronic internet grows, with its clubs, bulletin boards, and various fanciful "meeting places," the diversity of the real world will be mirrored in cyberspace. And as access to interactive television channels grows to the projected number of five hundred or more, the feared domination by a few major networks will be less likely. Thus will technology make it possible for individuals of similar temperament and taste to gather together, so assuring that all facets of the human spirit will continue not only to find expression, but to flourish.

The system does not work smoothly, to be sure, and there is no guarantee that we are heading for the best of all possible worlds. But there is more than enough evidence with which to refute the pessimistic assumption that in a technological world diversity is doomed. The ultimate lack of variety, of course, is to be found in a life *without* technology, where human potentialities are constrained by the demands of brute survival.

## THE TECHNOLOGICAL TOURIST

Even if the diversity of cultures is preserved, what are we to say about those remote scarcely inhabited portions of the globe that are now being penetrated by tourists? Thanks to technology, aren't we on the verge of destroying some of the most precious wildernesses left on earth?

"Exceptional Experiences for Selective Travelers!" So promises an advertisement in my college alumni magazine. Trips are scheduled to places that are "mysterious" (central Asia), "enigmatic" (Borneo), "primitive" (the Amazon), "compelling" (New Guinea), "exciting" (Tanzania), and much more. It seems that Paris, Rome, and the English countryside are old hat. The exotic trip, promising adventure spiced with a touch of erudition, has become big business. Experiences that once were reserved for the special few are now available to anyone with the time and money to take a two-week vacation abroad. (While some of these trips are quite expensive, others—featuring various degrees of "roughing it"—are no more costly than a typical European jaunt.) This extraordinary cultural change has been made possible by the evolving technology of travel, mainly by the proliferation of the jumbo jet. Once it became possible to transport several hundred people across the ocean en masse in a few hours, the consequences might have been foreseen. But jet planes are only part of the story.

My wife and I, succumbing to the blandishments just described, spent two recent holidays in places that until lately we would only have dreamed of going: the Arctic and equatorial Africa. We traveled north from Norway in a ship, one of several "explorer" type vessels that make a specialty of taking tourists to the two polar seas. Our comfort was assured by an excellent stabilizer system, and our peace of mind enhanced by the latest in navigational equipment. On a visit to the bridge I was enthralled to see our position on the globe digitally displayed, changing with each nautical mile, as electronic signals bounced off an orbiting satellite. With radar scanning the sea around us, sonar checking the depths below, weather reports streaming in from all about, and constant communication with the coast guard and other benevolent support institutions, one begins to feel almost snug. In Kenya we flew from place to place in small airplanes, not exactly futuristic, but also well endowed with modern equipment. For "game runs" we traveled in Toyota vans, with comfortable seats, roofs that popped up for viewing and photography, and suspension systems that were little short of miraculous.

On both trips the food was more than acceptable, thanks to refrigeration and various technologies of food processing. Synthetic fabrics and insulation kept us warm and dry in the north; improved "safari" clothing helped keep us comfortable in the tropics. We were amply instructed in how to cope with the climate, what to eat and drink, and more important, what not to eat and drink. In Africa, insects were kept at bay by chemical sprays, and on the off chance that we might get bitten, we were prepared with ointments and fortified with pills and shots. Aboard ship there was a physician and a pharmacy. In Africa we were assured of protection by an organization called the Flying Doctors. In places where not many years ago the slightest mishap could lead to disaster and death, we were remarkably safe and comfortable.

The people who operate such trips, making use of the technologies available, have developed the care of tourists to a fine art. A testimony to their success is the number of elderly people who are now traveling to the ends of the earth. To anyone who enjoys coming upon examples of the indominitable human spirit, meeting such senior citizens is a bracing experience. Yet, their presence in large numbers confronts one with proof that the trip is less demanding, less *adventurous,* than one might have supposed when dreaming about it in an armchair at home. I will never forget the thrill of going by rubber raft across an icy sea, surrounded by glaciers, a mere six hundred miles from the North Pole, and then the strange feeling of seeing an eighty-five-year-old lady picked up bodily by one of the ship's crew and deposited by my side upon the shore.

Is this development in tourism to be deplored, an instance of technology reducing the quotient of excitement and romance in the world? I think not. For the traveler who seeks new experiences without undue rigors, these trips are truly marvelous. It is a thrill to walk upon a glacier, or to spy a cheetah in the tall grass, no matter how safe one may be at the moment. (Also, let us recognize that at least *some* element of daring is involved: it could not be fun to be seriously ill on rough seas halfway between Spitzbergen and Iceland; and in Africa every now and then a tourist does get killed by a lion!)

We need not feel sorry for true explorers, since there are still many frontiers on land and in the oceans. As for the adventurous tourist, there is ample room—most of the world, in fact—where one can leave comfort and support systems behind and take one's chances with challenging climes and cultures.

If pure danger is one's fancy, then technology gives at least as much as it takes away. Daredevils who would have liked to chance running into an iceberg in the Arctic Sea, or being frozen in for the winter without hope of rescue, can try skiing down avalanche-prone slopes accessible only by helicopter, or hang gliding, parachuting, or some other newfangled hazardous activity. And if travel to exotic-sounding places is not as good for one-upmanship as it used to be, well, snobs will not fail to find substitutes.

More troublesome is the question of whether this intrusion by large numbers of tourists into remote places will not do serious environmental damage. Here, again, I think that the answer is no, although a qualified no it must be. Many of these trips are sponsored by scientific organizations, for people with at least an embryonic interest in the natural world. Such travel provides a breeding ground for environmentalists. One cannot see whales, or rhinos, and not become concerned about the precarious fate of these great beasts. One cannot delight in flora and fauna, and wondrous landscapes, oceans and rivers, without wanting to protect them.

In some cases the intrusion of tourists is little less than the salvation of the local biosystem. This is certainly true of the game parks of Kenya. These vast preserves could not be maintained if it were not for the travel industry. The local inhabitants are mainly interested in expanding agricultural cultivation, or in the case of such tribes as the Masai, in extending the feeding range for their cattle. Only a strict policy of conservation stands between such forces and the extinction of the wild game, and tourism is what makes this politically viable. Of course, the number of tourists must be controlled, or at least wisely distributed. Short-sighted politicians who want more tourist dollars immediately, by whatever means, are rightly challenged by those who would act more deliberately lest the popular parks be overwhelmed and the enterprise ruined by its own success. The animals also must be controlled. For example, in some areas elephants are destroying vegetation faster than it can regenerate; without intervention by scientifically trained experts, disaster will result. Nature is not always kind, nor is technology necessarily the enemy of nature.

The term *ecotourism* has lately come into vogue. From the rain forests of the Amazon to the frozen reaches of the polar regions, travelers can be found looking, learning—many of them strengthening their interest in environmentalism—and, of course, all of them paying, helping to support the habitat through which they roam. I see a headline in the travel section of the *New York Times:* "Ecotourism: Can It Protect the Planet?"[12] The answer is that in many parts of the world it has already helped.

When, more than a century ago, lovers of the wilderness protested the invasion of forests by the railroad, Edward Everett, a sage of the day, acknowledged that they had a point. "But gracious heavens!, sir," he went on to say, "how many of those verdant cathedral arches, entwined by the hand of God in our pathless woods, are opened for the first time since the creation of the world to the grateful worship of men . . . !"[13] To which many a tourist would say, amen.

If the worshipful tourist—the technological tourist—also turns out to be the salvation of the wilderness, then truly we have the best of all possible circumstances.

## FINE FEATHERED FRIENDS

There are many naturalists who are not so sanguine about the coexistence of wilderness and civilization. Among the most melancholic and apprehensive are members of the birdwatching community. It seems that there are fewer birds than there used to be, and this, say the birders, is very bad news. According to an editorial in *American Birds,* "the evidence is in and clear that several species are declining and many populations are showing signs of distress."[14] This is bad news for everyone, not just birdwatchers, because, according to this same editorial, the number and diversity of birds "mirror the health of the global environment."

A generation ago, Rachel Carson, in *Silent Spring,* used the distress of birds to alert the public to the dangers of chemical pesticides. Her warning had its good effect, and with the banning of DDT, the affected species, most notably the osprey and other raptors, substantially recovered. The present crisis, however, is more widespread and can't be traced to a single cause. The main problem seems to be "development"—development in North America where many species nest and breed, in South America where many species spend the North American winter, and on the flyways between the two.

I am a sometime member of the birding community and I share the concerns of my fellow birders. Yet I do not share the pessimism verging on angst that the crisis seems to have evoked. To the contrary, the history of birds in America gives me reason to be hopeful.

It is a fact—less known than I think it ought to be—that human development has serendipitously helped many species of birds to thrive and increase in number. The primeval forest, for all its enchantment, is not a congenial environment for most birds. When timberland is partly cleared, the "edge" that is created, with plants in various stages of growth, provides habitat for our most cherished songbirds. There are, according to Roger Tory Peterson, renowned dean of birding, perhaps a billion or two more songbirds in America today than there were before the arrival of the Pilgrims![15]

Many birds actually prefer living with humans to living in the wild, and I do not refer just to pigeons, English sparrows, and starlings. The list of species who have taken happily to towns and suburbs includes chimney swifts, phoebes, swallows, and nighthawks (who lay their eggs on the flat gravel roofs of commercial buildings).

This is not to say that human civilization is not a serious menace to our avifauna. Since the arrival of Europeans in North America five centuries ago, four native species have become extinct: the Labrador Duck, the Great Auk, the Passenger Pigeon, and the Carolina Parakeet, and probably three others that have not been sighted in over thirty years: the Ivory-billed Woodpecker, Bachman's Warbler, and the Eskimo Curlew. Add five American *sub*-species, including the Dusky Seaside Sparrow, the last of which died in captivity in 1987, and you still have a number that seems to me to be amazingly small. Any loss at all is tragic, but considering the speed and voracity with which the United States was developed, one looks for reasons why the destruction was not many times worse.

We could have exterminated most of our wild birds—for food, for feathers, for sport, for egg and nest collections, or through destruction of habitat—but we haven't. According to Peterson, "We seem to sober up at the eleventh hour, so we establish laws, game regulations, soil conservation practices, national forests, national parks, sanctuaries, and wildlife refuges."[16]

The first protective regulations date back more than three hundred years to the initial Dutch settlements. By 1776, twelve of the thirteen original colonies had some sort of game laws. In 1900 the National Audubon Society started to set aside bird sanctuaries, and in 1903 President Theodore Roosevelt signed an executive order establishing the first federal bird reservations. Today there are more than 350 of these reservations covering more than 20 million acres, managed by the United States Fish and Wildlife Service. The United States National Park Service administers over 357 parks covering more than 80 million acres. The national forests embrace 176 million acres. There are countless state, municipal, and private parks and sanctuaries. The hunting of waterfowl is strictly regulated, and the harming of most other wild birds is totally prohibited. Endangered species are safeguarded, however tenuously, by law.

In response to the current alarm about declining avian populations, the Audubon Society has launched a program called Birds in the Balance. Fact-gathering and study will be followed by actions to preserve habitat where it is most needed, including foreign lands. In anticipation of debate about the Endangered Species Act, the Interior Department has proposed a national biological survey to map the nation's ecosystems and biological diversity, much as the United States Geological Survey maps its geology and geography. In many localities, developers are negotiating agreements with government and environmental groups. I find these purposeful actions—past, present, and contemplated—tremendously reassuring.

In the end, we learn from the birds that nature is resilient but only up to a point. We also learn that our forebears treated the environment more sensitively than we might have thought, that we have reason to be thankful to them, and that we had better go and do likewise.

## 20/20 HINDSIGHT

Of course, our forebears did not always make good decisions, and as we plan for the future it is useful to remember the lessons of the past. This has become a platitude of discourse. Equally true, although less often remarked: If we become obsessed with the past we might succumb to paralysis. If we indulge excessively in hindsight—according to the saying, it is always 20/20—we might even learn the wrong lessons.

Take, for example, the tragic events that have followed from the use of asbestos. The mineral has excellent insulating qualities and for many years was widely used in buildings, ships, and various manufactured products. Now, however, asbestos is recognized as carcinogenic when inhaled, and exposure to it is stringently limited by government regulations. Everybody agrees that sound policies must be developed for removing or abating the material wherever people might come in contact with it, particularly in buildings. Also, attempts must be made to compensate the many tens of thousands of people—mostly workers who used the material in their daily jobs—for the devastation of illness and premature death. For a while this was done through individual lawsuits. After Johns Manville, the largest manufacturer of asbestos products, declared bankruptcy in 1982, the courts began to consolidate many of the cases, and funds for payment were established by corporations, receivers in bankruptcy, and insurance companies. The sad and costly process will eventually dwindle to a conclusion, although it is clear that adequate restitution can never be made.

As for lessons to be learned, attention is now turning to fiberglass and other manufactured mineral fibers that have long been deemed relatively safe substitutes for asbestos. Producers of these synthetic materials are undertaking costly new studies on the health effects of their products and are engaged in continuing dialogues with federal regulators. Beyond such steps, and being alert to protect the public from asbestos wherever it may be found, it is difficult to know what else can be done. We must do what we can to be fair and humane in dealing with those who have suffered, and we must take care not to repeat our errors. Life must go on.

There are people, however, who see the asbestos story in terms of good and evil. They accuse business executives and engineers, who in past years controlled the asbestos companies, of being sinful perpetrators of terrible crimes. According to this view, the lessons to be learned go far beyond how mistakes were made and future mistakes might be averted.

As far back as the 1930s some researchers and industrialists were aware of the health hazards to which asbestos workers were exposed. Liability lawyers have charged that, in the light of such knowledge, failure to take appropriate steps to protect the workers constituted "outrageous and willful misconduct." To the extent that this charge allows judges and juries to hold corporations culpable so that restitution can be made, I find no fault with it. But, taken literally, it leads to misplaced hatred of industry and its technicians; it leads to a bitterness that corrodes our already fragile social structure.

Paul Brodeur chronicled the asbestos story in a series of articles in the *New Yorker* magazine, later included in a book entitled *Outrageous Misconduct.* His accusations were harsh and unforgiving. But Brodeur's angry denunciation of corporate America contains evidence that militates against his presumption of wickedness. For example, he quotes a high-ranking industry executive (who was *himself* to die from asbestos-induced mesothelioma): "I realized that precautions needed to be taken in the handling of this material. What I realize now is I was in error about the extent of precautions which needed to be taken."[17]

An even more telling argument against the "misconduct" theory is found in Brodeur's description of a symposium held at Saranac Laboratory in 1952. This event brought together doctors (more than fifty), research scientists, academics, lawyers, public-health officials, insurance company executives, and journalists, as well as people from the asbestos industry. Ample evidence was presented linking asbestos to cancer. Yet more than a quarter century was to pass before society came to grips with the problem in a definitive way. Clearly something other than evil intent was involved. Nor is it rational to condemn, as does Brodeur, "a huge cross-section of the institutions that make up the private-enterprise system,"[18] including, incidentally, trade unions. This implies that socialist nations were more prescient and caring about public safety, which we know they were not.

The plain fact is that our communal view of the world changes from time to time, not only about religion, sex, civil rights, and feminism, but also about environmental pollution and worker safety. I can remember that when I first entered the construction industry, work sites were dirtier, noisier, and very much more dangerous than they are today. One heard it said that for every million dollars spent there would be a worker's life lost—a rule of thumb that speaks volumes about the ethos of the times. People expected less protection than we do today: from injury, sickness, and poverty. But dare we conclude that this made them—or even the leaders among them—morally inferior? Such hindsight moral egotism would lead us to despise Thomas Jefferson for tolerating slavery, indeed, to despise all people who lived in earlier times accommodating themselves to public executions, bearbaiting, and the exploitation of child labor.

If we have gradually developed higher standards of public health and worker safety, we should be proud of our moral growth but also realistic. More stringent precepts have evolved in large part because we can now *afford* them. The wealth that allows us to be more caring and sensitive—as well as the new knowledge that enables us to reduce risks—comes from the very technological development that it pleases so many second-guessers to view with contempt.

The asbestos disaster challenges us to cope effectively with the dangers at hand; to compensate, humanely and rationally, those who suffer from past mistakes; and to develop materials and methods that will safely and efficiently do what asbestos used to do. Insulation, a key element in energy conservation, has a compelling claim on our attention.

As for moral outrage, let us direct it against the inequities of our own time, of which there is no shortage, instead of against the sins and foibles of people who lived a generation and more ago.

In this chapter I have suggested that technological progress need not mean the end of many things we hold dear. If we avoid the bleak modernity of *Star Trek,* resist homogenizing the world, protect our flora and fauna, and refrain from tearing our society apart with recriminations, we will have accomplished much, but mostly by way of defense. Let us now switch to an affirmative mode. What new prospects and worthy goals lie before us in the engineered world of the future?

CHAPTER 3

# BRIGHT ASPECTS OF TECHNOLOGY

## THE QUEST FOR COMFORT

I am grateful for my pillow. Yes, my pillow, that commonplace object upon which I rest my head through the night. It is made of a spongy plastic that provides just the right combination of support and pliability. I have experimented with many pillows and chosen the one that suits me best. I do not take it for granted, this manufactured article that gives me so much comfort. A lot of people worked hard and ingeniously over many years to create it. My wife prefers a softer pillow filled with down. I appreciate living in a world where one can choose just the kind of pillow that feels best. I relish our sheets and blankets, too, along with the mattresses and box springs. What marvels of linen-growing, wool-shearing, fiber-making, weaving, manufacture, and transport have been accomplished for our benefit! Some mornings, as I open my eyes—warm, safe, at ease—I am suffused with gratitude for human ingenuity and enterprise, with reverence for the achievements of technologists.

This is not to say that my primary concerns relate to comfort and the devices that effect it. Along with Walt Whitman I give thanks "For health, the midday sun, the impalpable air—for life, mere life . . ." I agree with Walter Scott's minstrel that "love is heaven and heaven is love." But the poets fail to do justice to God's world when they ignore, or even undervalue, the handicraft impulse that has evolved into engineering. *Technology* is a rather bleak word, associated in the public mind with smokestacks, dynamos, and parking lots, and lately with electronic devices. It deserves a better fate, symbolizing as it does so many of the splendors and potentialities of human creativity.

Technology begins with the quest for comfort. Humans, like all living creatures, seek sustenance and safety. The maintenance of life is the priority that life sets for itself, that has been bred into it by the exigencies of evolution. And as comfort is a symptom of survival—consisting mainly of those feelings that go with being fed and protected—comfort is something we pursue. We seek it largely through technical activity, by making things that keep us warm, dry, nourished, and safe. So the technological impulse—and the appreciation of technology—are intrinsically related to comfort.

Of course, the thought of too much comfort makes us uneasy. This stems from a cautionary lesson taught by evolution and embellished by philosophy. The hardiness of primitive hunters and warriors is part of our nature. We believe in the Spartan virtues, and value the rigors of exercise and harsh climate. There is pleasure to be found in camping out, in sleeping under the stars on the hard ground. There is even a certain appeal in danger. We disdain excesses of luxury (although it is not always easy to define where appropriate desires end and excessive luxuries begin). Nevertheless, the claims of comfort are many and elemental, as are our ties to the technology by which we try to achieve it. If it is sinful to worship comfort, it is equally sinful to take it for granted, and to fail to appreciate the ingenuity and effort by which it is achieved. Further, it is wrong to ignore the poverty and destitution that characterize the lives of so many of our fellow humans, and to neglect the skill and effort that is the only means of mitigating so much misery. What could be a more noble goal than to secure a modicum of comfort for the masses whose daily lot is suffering?

Comfort may be the beginning of my appreciation of technology, but it is by no means the major component. The human spirit has, in its lust for life and its consequent quest for comfort, become amazingly clever, dexterous, intelligent, and inventive. The *doing* of engineering, and the contemplation of that doing, is even more awe-inspiring than the contemplation of engineered products.

Fond as I am of my pillow, I am perhaps even more fond of my electric alarm clock. It is so complex and yet so simple. In summoning me to a new day, it speaks to me of energy and purpose, of missions and objectives. It tells me what time it is, to the minute. There are sages, I know, who deplore the awareness of time that is so much the mark of our modern age, who believe that life has been getting worse ever since some nameless genius of the late fourteenth century perfected the mechanical clock. Lewis Mumford, for example, regrets that with the coming of timekeeping, "Eternity ceased gradually to serve as the measure and focus of human actions." He concludes rather dourly that the clock "is the key-machine of the modern industrial age."[1] Perhaps he is right, but if so I find no cause for lamentation. I am a product of the modern industrial age. Technology is my birthright, and I revel in it.

When I was a boy I had a classic round Westclox with a bell on top. I enjoyed winding it every day and thinking about the complex mechanism that made it go. But the alarm was loud and jarring, and occasionally the ticking could be an annoyance. So when electric timekeepers came along, and rapidly dropped in price, I resolved to make the change. Through the years I tried various types, including a clock radio, but eventually settled on a plain low-slung model that wakes me with an agreeable dinging sound, and displays the time with glowing red numbers that I can read in the dark without my glasses. I've heard it said that a digital display is somehow less desirable than a dial, but I don't find this proposition convincing. I like numbers and find their configuration pleasing, even when they tell me that I have awakened inadvertently at some inauspicious hour. The numbers speak to me in a universal language, denoting quantity and proportion, or kindling thoughts of other people and locales. When it is 3:00 A.M. in New York, it is just midnight in California, and mid-afternoon in China; sometimes I wonder what is happening in those remote places while I turn over and puff up that wonderfully comfortable pillow.

My mornings, needless to say, do not typically begin with an ode to technology. Like most people—engineers included—personal concerns prevail over cosmic reflections. But on some days—

some special days—an awareness of my pillow or my alarm clock or some other manufactured object, will start me thinking about the wonders of technology. And then there is no shortage of reasons to reflect, marvel, and be grateful.

Viewed in a philosophical context, the engineering endeavor can be seen to contain elements of the sublime. This is true even though—or possibly partly because—it begins with the primitive search for comfort.

## THOUGHTS ALONG THE NILE

Beyond survival, beyond comfort, beyond individual aspirations, there lie those vast communal undertakings that redeem engineering even in the eyes of its severest critics. In the Western world the medieval cathedrals are probably the outstanding example, followed by a variety of monuments, shrines, parks, museums, libraries, and so forth. However, looking back over the total sweep of human history, the most striking demonstration of public enterprise—considering age, immensity, and state of preservation—is to be found in the monuments of ancient Egypt.

An engineer traveling in Egypt cannot help feeling like a pilgrim returning to an ancestral shrine. That fascinating land has long attracted many visitors; but for an engineer it has a special resonance. At least that is the way I felt when I had an opportunity to visit there.

This is where it all began, I mused, as our riverboat drifted downstream with the Nile current. This is where humans learned to join irrigation canals into large networks, to build dikes and watergates, to work cooperatively in creating agricultural abundance for a large population. The energy thus liberated was then applied to building some of the most monumental structures the world has ever seen. This is where engineering was born, and with it the beginnings of civilization as we know it.

Of course, many engineering techniques originated in places remote from the Nile Valley. But if we think of engineering as a group activity, with an important social and political component, then surely Egypt can claim a special priority. More than five thousand years ago, along seven hundred miles of the fabled river, the Egyptians were united under a single ruler, and engaged in hydraulic engineering on a massive scale.

I found the famous antiquities—the pyramids, the sphinx, the temples at Karnak, and many others—every bit as awe-inspiring as the travel books say they are. I know that some chroniclers have been disenchanted by the notion that these mighty works were built by slaves and inspired by the dread of death. Yet, according to many of today's Egyptologists, such presumptions have been erroneous. During the months when the river was in flood, making it impossible to perform agricultural chores, large numbers of people were available to work on construction projects. Consequently, the tombs and shrines were the public works of their day, built mostly by agricultural laborers in the off-season. As for macabre theories about crypts and burial artifacts, historians are also having second thoughts. They suggest that the Egyptian interest in an afterlife stemmed not from fear and anguish as in many cultures, but rather from a zestful fondness for life itself.[2] Be this as it may, the splendid structures bespeak a radiance of spirit that must enchant the visiting engineer.

Moreover, I was pleased to find that some of the engineers responsible for the great works are known to history and accorded full credit. Around 2950 B.C., Imhotep, prime minister to the pharaoh Zoser, and the first engineer whose name is recorded, built the famous step pyramid at Sakkara. Centuries later the Egyptians elevated him to their pantheon of gods, certainly a unique fate for a member of the engineering profession. In a tomb at Abydos, dating to about 2500 B.C., it is inscribed that a hydraulic specialist known as Uni was "Superintendent of the Irrigated Lands of the King." Grandest of all, perhaps, was Ineni, "chief of all works in Karnak," and "foreman of the foremen." He left a striking message for posterity: "I became great beyond words," he wrote. "I did no wrong whatsoever."[3]

No engineer today would dare to think in such terms, least of all—ironically—in Egypt. After the thrill of the antiquities comes the recognition that there is almost as much in this ancient land to abash engineers as there is to inspire them. Certainly the High Dam at Aswan is a classic case of technology gone awry. Built with the worthy goal of providing power and controlling river floods, it is silting up with alarming speed, has inundated precious lands in Nubia, and wreaks havoc with the ecology of the Nile Delta.[4] Also, as the water table rises in the sandy soil downriver from the dam, moisture seeps into the precious ancient monuments, threatening serious damage.

Fortunately, corrective action is being taken. Say what one will about technological fixes, it is gratifying to see silt being removed from the upstream side of the dam and spread over desert to create new cultivatable land. And surely one of the highlights of an engineer's trip has to be the temples of Abu Simbel, salvaged in 1963 and rebuilt on high ground as the impounded waters rose behind the new dam. Moisture in the monuments is a difficult problem, but it, too, is being addressed with schemes for drainage and protective coatings.

The ancient Egyptians, stirred by a primal impulse, grappled with nature in pursuit of material abundance. Having succeeded on a scale previously unequalled, they were inspired to go beyond physical need and gratification, to create temples and majestic works of art. This manifestation of the human spirit still endures. We create engineering works, not only to survive and prosper, but also to express transcendent aspirations.

The same spark that inspires us to create such wonders should motivate us to preserve them. Thus one hopes that today's Egyptians—with the help of the world's engineers—will conserve the antiquities that testify so movingly to human potential.

## FORWARD INTO THE PAST

It is good to preserve precious antiquities. It is bad to harm them. It is also good to unearth new evidence of vanished civilizations and their arts. There is fear that as we move into the engineered world of the future we are doing irredeemable harm to our ancient heritage. But the truth is much more complex. It may be that discovering, conserving, and interpreting artifacts of the past is one of the noblest and most creative technological enterprises of all. It adds not only to knowledge of the past but also to the quotient of beauty in the world.

When a sunken Spanish galleon, loaded with treasure, was discovered off the Florida Keys in the summer of 1985, the news created considerable public excitement. Stacks of silver ingots brought up from the bottom of the sea; wooden chests spilling over with gold coins—[illegible]les our national psyche, nourished as it is on get-ri[illegible] [illegible]tteries, and TV game shows. Also, the trea[illegible] [illegible]e of adventure and our admiration [illegible]s and the derring-do were not [illegible] [illegible]heological significance [illegible] [illegible]usiastic reactions [illegible]

[illegible] spectacular find," said [illegible] [illegible]heology at Texas A&M [illegible] [illegible] an intact cargo that can [illegible] [illegible]ation in the New World." [illegible] [illegible]ologist, with forgiveable hyperbo[illegible] [illegible]n enormous time capsule, as important as [illegible] [illegible]ng Tut's tomb."[5]

As an engin[illegible] [illegible]yself thinking about the technologies that helped mak[illegible] discovery possible—particularly side-scanning sonar, scuba [illegible]ving gear, and a high-speed magnetometer—technologies that were not available to treasure hunters of earlier times. I was especially intrigued that so much newly developed technical equipment was being used to recover the past.

Engineering progress is usually associated with the future. In the slogans of engineering celebrations and the titles of engineering meetings the dominant theme is usually along the lines of "the world of tomorrow" or "the years ahead." Charles Franklin Kettering, who for thirty years was head of the G.M. Research Corporation, expressed a typical engineer's view when he said, "We should all be concerned about the future because we will have to spend the rest of our lives there." Who would have guessed that one of the most interesting things about the future will be the knowledge we will gain about the past?

The bottoms of the seas contain an incredible wealth of information about the past, most of which has heretofore been beyond our reach. In 1960, Swiss engineer Jacques Piccard and U.S. Navy lieutenant Don Walsh dove the navy bathyscape *Trieste* 35,800 feet down into the Mariana Trench, the deepest spot in the ocean. But, soon after, that unwieldly steel sphere was retired from duty, and for more than three decades such remote regions were beyond human reach. Because manned submersible craft were not considered safe beyond a depth of about 12,000 feet, half the ocean floor remained hidden from view. However, a new generation of manned vehicles is now able to travel 20,000 feet below the surface, while unmanned robots are capable of reaching the nethermost abysses.

These robots, known as remotely operated vehicles (ROVs), are already in wide use, not only locating and exploring sunken ships, but also working on offshore oil installations, undersea cables, military reconnaissance, and scientific exploration of the ocean floor. Although they accomplish marvelous feats, ROVs need to be tethered to support ships by umbilical cables that relay control signals, power, images, and other sensor data. This presents problems of maneuverability; also economics, as the support ships entail significant cost. Accordingly, much effort is being devoted to developing autonomous underwater vehicles (AUVs), which soon will roam the oceans unattended. There are formidable technical problems: Control cannot be achieved by radio waves, and navigation is especially difficult in an often featureless medium. But between 1987 and 1995, twenty AUVs were developed by seven different nations, and all signs point to increased activity in this fascinating technological realm. The principal motivation may be financial and military, but the benefits for art and history are marvelous to contemplate.

The surface of the earth, of course, also contains archeological treasures that we have not even begun to explore. Since the 1940s, the historical significance of discoveries has been greatly enhanced by dating techniques that utilize the radioactive decay periods of certain elements, notably carbon 14. In the 1990s, the development of aerial photographic techniques initiated a new and equally exciting research era. Airborne instruments, sensitive to parts of the electromagnetic spectrum invisible to the eye, exposed patterns of ancient life in the Chace Canyon of New Mexico. Buried prehistoric walls, buildings, agricultural fields, and roads were detected that long predate the well-known cliff dwellings. Prehistoric footpaths, long blanketed by volcanic ash, have been identified in Costa Rica. A Mayan causeway running through the jungle is one of several ruins located in Mexico and Guatemala.

Several years ago, a system called Space Imaging Radar, carried on a space shuttle, peered underneath the deserts of Egypt. The data recorded was combined with optical sensing information from satellites, producing digital images that were then enhanced by computer manipulation to bring out subtle details. Thus ancient caravan routes were revealed. Following through with a land expedition, and using sounding devices and on-site ground-penetrating radar, archaeologists, in 1992, uncovered the fabled Arabian city of Ubar. The same techniques have been used to map ruins along the legendary Silk Road in the desert of northwestern China. The prospects for future discoveries are nothing less than thrilling.

Much has been written of how airborne pollutants damage works of art; but we would do well to remember how chemical science is applied to the cleaning and restoring of paintings. In many museums and libraries, humidity and temperature are controlled by sophisticated systems forestalling a decay that otherwise would be "natural." Photography and other reproduction techniques make widely available priceless works, from cave paintings to medieval illuminated manuscripts.

The use of electronics for the benefit of art appears to have no discernable limits. Standard equipment for museums today includes scanning electron microscopy (detects microbes on canvas), polarized light microscopy (helps identify pigments), infrared reflectography (reveals underdrawing), ultraviolet light (reveals inpainting), X-ray radiography (shows locations of heavier elements such as lead and vermilion), X-ray diffraction (detects compounds through characteristic crystalline structure), and electron beam microprobe (determines which elements are present in a sample and in what quantities). Major institutions also apply mass spectroscopy (analyzes lead isotope ratios in pigments) and high-pressure liquid chromatography (analyzes binding media such as oil and glue). A handful of institutions utilize energy-dispersive X-ray fluorescence (detects elements in a painting without removing particles of pigment). And a few museums, using computers, have image processing capability (enhances faint details; clarifies underlying paintings; allows restoration that is not immutable). Where a cyclotron is handy, or a nuclear reactor, additional analytical techniques can be employed.

Capabilities for analysis, preservation, and restoration have reached such scope that disputes have arisen between proponents of "make it as it was," "save it as it is," and "leave it to age as it will." Was it a mistake to restore Michelangelo's paintings in the Sistine Chapel to what is thought to be their original brilliance? Should the most precious statues of Florence be moved indoors or treated with protective coating? Is the cleaning of buildings at Angkor Wat in Cambodia being done with too harsh a detergent, removing details with the dirt? Is it appropriate for a Japanese crane company to lift fallen statues on Easter Island? These arguments, vexing as they may be, are surely better than the hand-wringing over loss that prevailed in the arts world not too long ago.

The conflict between destructive and restorative aspects of technology can be seen perhaps most clearly in Venice. There airborne acids from nearby industries are eroding statues and buildings while at the same time scientifically trained curators ply their healing crafts; and new tide gates are designed to keep stormy seas from destroying the city altogether.

Will technology discover and preserve more than it destroys? As I perceive the marvels of each new day, I am inclined to think so. However, as in so many matters, it depends upon what we as a society decide to do. The preservation of Venice, like all art patronage, requires the appropriate allocation of resources, that is, the spending of money. If we pursue the past only when the prize consists of gold and silver booty, the net result will surely be a loss. Our quest must go far beyond treasure-laden wrecks on the ocean floor. Indeed, the Spanish ship with which I started this section provides a striking moral: A civilization that puts too high a value on the pursuit of gold—a short-sighted, noncreative, *non-technological* activity that obsessed Spain in the sixteenth and seventeenth centuries—is not destined to survive the tides of history.

## TEARING DOWN AND BUILDING UP

Related to the question of what we do with antiquities is the question of how we treat the aging artifacts of our own era. In our need and desire to move ahead do we recklessly build and destroy and then build again in a cycle that is wasteful, not only of resources, but also of cultural heritage? Certainly this is the perception in some quarters.

For example, New York newspapers report from time to time, with a mixture of excitement and alarm, that the owners of Madison Square Garden are thinking of tearing down that famed edifice and building a replacement arena farther west in Manhattan above a railroad storage yard. This would allow the present Garden site to be used for a large office complex. The prospect is somewhat unsettling, as the building that would be demolished was constructed not so very long ago (1968). The "old" Garden, which stood from 1925 to 1968, was razed because it was ugly and outmoded, not up to "modern" arena standards. Its two predecessors had been replaced in turn because they were deemed too small. To demolish structures that are sound and useful seems "wasteful"—at least that is a word I have heard used since the proposal was first announced. The New York Coliseum—built only a decade before the Garden, but perhaps made superfluous by a larger Convention Center—is also scheduled for demolition, as are hundreds of other solid buildings across the country.

When multimillion dollar structures are torn down as if they were shacks, when strong walls, watertight roofs, and efficient heating plants are purposefully turned into trash, one's first conclusion is that we have indeed become the ultimate "throwaway" society. How should we regard this strange phenomenon? What judgment should we pass? What action should we take?

There are two main considerations, it seems to me. The first is practical, based on an evaluation of our resources; in other words, seeking to act with long-term economic prudence. The second is aesthetic and moral, choosing the course of action that is worthy of a civilized people, that satisfies our communal aspirations.

Buildings are big, and tearing them down is visible and dramatic, yet not, in the scheme of things, as wasteful as it might appear. We have plenty of sand, cement and gypsum, iron ore for steel and clay for bricks. Those building materials that have significant and lasting value—copper, for example—are salvaged during demolition. There is no shortage of workers in the building trades. Indeed, as manufacturing becomes increasingly automated, and we change into a white-collar and pink-collar "service" economy, a robust construction industry becomes a social and economic asset.

Thriftiness is a virtue and yet we are not inherently a thrifty people. We cut down forests to provide newspapers, magazines, and books, a process in which we take much pride. (Particularly as we attend to reforestation.) We use and discard, or recycle, clothing, packaging, and a thousand other familiar items—so why not buildings?

In this fast-moving world, changing real estate prices can often make brick and mortar utility a secondary consideration. *Value,* however we interpret the term, is surely related to optimum use. Also, buildings have a way of becoming obsolete, just like other products. Schools fall vacant as populations age or move. Hospitals are closed as the average length of patient stay decreases, or rebuilt as the nature of treatment changes. In London's financial district (the heart of conservative sentiment, one would think) a twenty-five-year old office tower is demolished and replaced because it doesn't have the fifteen-foot floor-to-floor heights its owner requires "for current banking technology."

Nor is alteration and renovation the ready answer to problems of building obsolescence. As a builder, I have done many studies of conversion projects—making apartments out of hospitals or offices out of schools—and more often than not I find that demolition and reconstruction is the most economical solution. This may be counterintuitive, but it is true. Rehabilitation has been oversold as an instrument of urban revitalization, at least from the point of view of building costs. The figures just do not work out the way idealistic planners would like them to. Where premature decay is the problem, rather than obsolescence, the sad truth is often the same. (Good maintenance is the secret of good health for buildings, just as conservation is the best source of energy savings, but unfortunately this message is hard to sell.)

No, the argument against building demolition cannot rest on economics. Waste is not the issue, at least when the market tells us it is not. If we want to save buildings, we must usually argue on aesthetic and moral grounds. In fact, we must be prepared to make economic *sacrifices* for the sake of communal felicity. This is the philosophical commitment that stands behind the architectural preservation movement.

In 1965 the City of New York enacted its Landmarks Preservation Law. This was done in the aftershock of the razing of Pennsylvania Station (ironically to make way for the Madison Square Garden now threatened with demolition). The law declared it a matter of public policy that certain buildings or neighborhoods should be protected from demolition or radical change. Protection is given to those buildings or districts deemed precious to the community. Selection is made by a commission consisting of three architects, a realtor, a city planner or landscape architect, a historian, and five other members, either laymen or additional professionals. There is due process, including public hearings and review by the City Council. In case of economic hardship, building owners may be granted tax abatement or other assistance, or in extreme cases, remission of the landmark designation. Owners are entitled to a return on their investment; but they may be required to forego the speculative profit that might otherwise dictate demolition.

The commission is at pains to point out that economic benefits often follow in the wake of good intentions. Revitalized and stabilized neighborhoods can stimulate property values; graceful and picturesque cityscapes can be good for business and tourism as well as for the soul. Private initiatives—such as the $50-million-dollar restoration of Carnegie Hall carried out in the mid-1980s—interact synergistically with the public commitment.

None of this has anything to do with Madison Square Garden, which is neither beautiful nor distinctive. Let it be torn down and rebuilt elsewhere if the market so dictates. This would be very much in tune with the American spirit. Hear Thomas Jefferson on the subject: "I think there is nothing so much in the world I like as tearing down and building up."[6]

An equally vital aspect of the American spirit is expressed in the slightly stilted but nevertheless lovely language of New York City's Landmarks Preservation Law: "It is hereby declared as a matter of public policy that the protection, enhancement, perpetuation and use of improvements and landscape features of special character or special historical or aesthetic interest or value is a public necessity and is required in the interest of the health, prosperity, safety and welfare of the people."

## HEGEL AND THE RAIN FOREST

"The essentially tragic fact," said the philosopher Hegel, "is not so much the war of good with evil as the war of good with good." In the engineered world this truth is brought home to us again and again. We try to do good and find that we are unwittingly damaging something that we hold dear. Yet such dilemmas can evoke noble impulses and creative thought. It is often in challenge that we find the ultimate expression of our humanity.

Consider the example of electric power in Brazil. The population is exploding, and in order to improve quality of life the government plans to expand electric generating capacity. Lacking significant reserves of oil or coal, and plagued with chronic technical difficulties at its single nuclear plant, the decision is made to build hydroelectric dams. My first reaction upon reading of this was to approve heartily. According to the environmentalist literature I have seen, hydroelectric power can be ecologically benign, far preferable to the burning of hydrocarbon fuels, which brings acid rain and the greenhouse effect, and better than nuclear power, which gives us safety problems, radioactive waste, and large doses of political angst. And where could it make more sense to exploit river resources than in Brazil, which within its borders contains one fifth of the world's fresh water? The Amazon runs four thousand miles to the Atlantic Ocean, and at least fifteen of its tributaries are more than one thousand miles long. Dams already provide 95 percent of Brazil's electricity, and the planning directors of the state power monopoly expect to maintain this proportion. One hundred thirty six prospective dam sites have been identified—more than half of them in the Amazon Basin—and at least ninety of these are on the planning agenda.

Energy without environmental cost. It seems too good to be true. And it *is* too good to be true. Complications arise because Brazil's thousands of miles of rivers run mostly through low-lying country. This means that building a dam entails flooding large areas of jungle. The Balbina Dam, for instance, built on the Uatuma River near Manaus, flooded nine hundred square miles of jungle to produce only 250 megawatts, barely half enough for the needs of the nearby city. The Samuel Dam, on the Jamari River near Porto Velho, uses land somewhat more efficiently, creating a lake of two hundred square miles and producing 217 megawatts. But it, too, is insufficient to meet local demand, much less contribute to the contemplated supply for the distant southern cities of São Paulo and Rio de Janeiro. In some parts of the Amazon Basin the land is so flat that long dikes have to be built in locations remote from the dams in order to keep the lakes within workable confines. The problem has been compounded because tree cover inhibits aerial topographical surveys, and erroneous assumptions have been made.

Looking at the Amazon Basin, an area as large as the United States east of the Mississippi, some Brazilians see little reason to be concerned about flooding a few thousand square miles of jungle. "There is plenty of space," said an engineer looking out across the Balbina Reservoir. "If one day we need the land, we can take away this dam."[7]

Environmentalists worldwide—including many Brazilians—do

not share this cavalier attitude. They perceive destruction of the Amazon Rain Forest as a potential disaster that would mean the extermination of thousands of plant and animal species and entail adverse effects on global climate. (Because trees absorb carbon dioxide and emit oxygen, they are a countervailing force to the greenhouse effect.) There are already great concerns about the burning of forest for agriculture, and the spread of mining and other industrial development. Flooding large areas for the sake of creating power—which in turn is likely to attract more development—only makes a bad situation worse.

In an effort to retard this process, concerned observers have called upon international creditors to cut off funds for new projects. The World Bank has demanded environmental safeguards as a condition for processing vital loans for energy development. Brazilians resent being pressured in this way, particularly by citizens of developed nations who didn't let environmental niceties get in the way of exploiting their *own* natural resources. The Brazilian government, joined by the seven other nations of the Amazon Pact, has denounced outside interference in the region's environmental affairs.

On the other hand, Brazilians themselves are looking for new and creative solutions. Some of their studies show that harvesting rain forest fruits, cocoa, and latex may be economically and socially more viable than flooding for power or large-scale clearing for other purposes. Local efforts are amplified as international consciousness-raising develops. The Global Environmental Facility, the Tropical Forestry Action Plan, the International Tropical Timber Association—these are just a few of the initiatives launched during the past decade.

We see that using hydroelectric dams to meet Brazil's energy needs—which at first seemed an ideal solution to a serious problem—confronts us with new difficulties, not only in environmental policy, but also in relations between industrialized nations and citizens of the Third World. Adding to the urgency of the problem is the fact that Brazil's population is growing at an alarmingly rapid rate. We can speak about the need to control birthrates, but in the meantime millions suffer. The tragic war of good with good. Hegel's words are made manifest in the depths of the Amazon jungle.

In this no-nonsense age of science and technology one rarely thinks of philosophers. Still, it is clear that the scientific approach has its limits. Systems analysis alone cannot tell us what to do about electric power in Brazil. Unless wisdom and sensitivity are brought to bear, the situation might escalate into confrontation.

According to Hegel, the cardinal principal of life is change, manifested in what he called the dialectic process. One concept (thesis) inevitably generates its opposite (antithesis), and the interaction of these leads to a new concept (synthesis). The new concept in turn becomes the thesis of a new triad. The process never ends. In every stage of history there is a contradiction that only "the strife of opposites" can resolve. This has certainly proved to be the case with economic development and environmental concern.

Hegel cannot help the Brazilians decide how many dams to build in the Amazon Basin. But his vision of life as a process of coping with opposites, and somehow reconciling them—and then starting again in never-ending struggle—encourages us to get on with the work that needs doing. This work calls for diplomacy as much as for science and technology, and yes, it calls for philosophy as well—supplemented, to be sure, by effective systems analysis. Some of our current technological predicaments are so perplexing that they force us to reflect on who we are and what we really want. Thus, as an unexpected side benefit, engineering progress leads us back to the most profound depths of philosophical speculation.

CHAPTER 4

# WAR, PEACE, AND FREEDOM

## TECHNOLOGY AND AGGRESSION

Of all the reproaches leveled against technology—and against engineers and engineering—the most grave relates to war. The connection between technology and war predates history. The origin of the engineering profession is tied to military needs, military sponsorship, and military schools.

Yet, without professing any unique virtue for themselves, engineers have good reason to argue that war elicited technology rather than the other way around. Fighting is inherent in human nature. Sigmund Freud concluded somewhat gloomily that "there is no likelihood of our being able to suppress humanity's aggressive tendencies."[1]

The instinct to make war coexists in the human breast with the impulse to be technological. It cannot be said that technology generated war, or even increased its importance in human affairs. Or, rather, it *is* said, but the facts do not support the charge.

When I visit a zoo, and see the monkeys and apes at play, or peacefully munching on fruits and greens, it seems perverse that humans evolved into hunters and warriors. But then I read of chimpanzees in the wild, hunting in packs, killing monkeys, antelope, and even wild pigs. "Drumming, barking, and screaming," as described by scientist observers in the Ivory Coast, the chimps seem eerily similar to a human war party. When the prey is sighted, "the hunters scatter, often silently, usually out of sight of one another but each aware of the others' positions."[2] Coordinated action is the key to their success. The luckless victim of the hunt is killed on the spot, typically by a bite to the head or neck, and torn apart for a shared feast. Although the chimps do not war with members of their own species, each group takes care to guard the borders of its own domain. During approximately monthly forays, "which seem to be for the purpose of territorial defense," the animals are "clearly on edge and on the lookout for trouble."[3] In the activities of these relatives of humanity, one senses the evolution of warlike instincts.

Even if all apes were congenitally peaceful, evidence shows that early hominids were hunters and sometime fighters. Killing with teeth and bare hands—and certainly with found objects—clearly preceded the manufacture of weapons. And some of the killing, like the hunting, was done by organized groups. War preceded technology. To be sure, hunting, and then war, evoked much technology, but degrees of savagery seem unrelated to levels of technical development.

Observing certain "primitive" tribes who live at peace with nature and with each other, one is inclined to presume that war, however widespread it may be and has been, is not an essential element of human life. We can pray that this is so. Yet the example of aboriginal societies, for all its appeal to some anthropologists, is far from reassuring. European explorers encountered peaceful tribes in various parts of the world, but they seem to have come upon at least an equal number for whom war was customary. Even Kirkpatrick Sale, author of *The Conquest of Paradise,* a fierce denunciation of Western culture, and a celebration of the native Americans of pre-Columbian times, admits that probably one third of the tribes practiced some form of war.[4] The question of whether early humans were *mainly* warlike or *mainly* peaceable will always be debated by experts, since definitive proofs cannot be advanced. But it is widely accepted that organized combat was an important element in earliest human life.

Archeological evidence confirms that warfare was conducted at least from the Late Palaeolithic onward. The earliest civilizations inherited from prehistoric ages a legacy of weapons development, fortifications, offensive and defensive strategies and tactics, and a sense of territoriality. "As soon as man learned how to write, he had wars to write about."[5]

The written historical record deals mainly with the rise and fall of empires as determined by military ventures. Wars have been practically ceaseless, as we know from the writings of Assyrians, Greeks, Romans, Turks . . . and just about every social group right up to modern times. During the sixteenth century, Spain and France were scarcely ever at peace; while during the seventeenth, the Ottoman Empire, the Austrian Habsburgs, and Sweden were at war for two years in every three, Spain for three years in every four, and Poland and Russia for four years in every five. Nor was warfare a special feature of the Western cultures with which Americans are most familiar. Chronicles of Asian history reveal the same bloody story. In China, for example, between 770 and 221 B.C., there were only seventeen years free of hostilities.[6]

Through all this carnage, the goal of technological superiority has been a continuing theme. Thinking of the carpet bombing of cities in World War II, and then the destruction of Hiroshima and Nagasaki with atomic bombs, one is tempted to postulate that in the age of high technology humans reached new depths of depravity. But history tells us of hundreds of cities that were razed and in which all occupants—tens of thousands at a time—were put to the sword. As late as 1994 we have seen in Rwanda butchery on a massive scale, most of it accomplished with machetes.

It is futile to seek a direct correlation between technological advance and brutality in war. One might rather observe that as weapons have become more advanced technologically, revulsion against war has increased, until finally, with the coming of the "ultimate" weapon, the determination to end war became an expressed objective of the world's peoples.

Of course, we are a long way from achieving that goal, as evidenced by battles that continue to rage in various parts of the world. There is ample cause for discouragement. Yet, during the 1991 Persian Gulf War there were developments that seemed totally new in human affairs and which I found heartening.

## THE NINTENDO WAR

While briefing the press during the final hours of the conflict against Iraq, General Norman Schwarzkopf told reporters not to forget that human lives were at stake. "This is not a Nintendo war," he said sternly.

Yet, as Desert Storm moved to its swift and stunning conclusion, I found myself slipping into a Nintendo frame of mind. I could not help being dazzled by what *Newsweek* called, "the boy-toy glow surrounding high-tech weapons."[7] Doubtless, being an engineer had something to do with my reaction. The ingenious contraptions worked, and a part of me responded with an instinctive "Wow!"

Of course, one didn't have to be an engineer to marvel at TV pictures of laser-guided bombs flying into air vents, nor to be awed by tales of cruise missiles guided to distant targets by internal computerized maps. There is a primal "Wow!" within us all. There is also a fierce and wary barbarian who wants to see the target hit.

Even after the frenzy of battle had passed, however, the Nintendo metaphor lingered on. There was a widespread feeling that the war had been a grand display of technical ingenuity, a triumph not only for American fighting forces, but also for American engineers. Yet, I thought, how can engineers take satisfaction in the victory without also feeling acute discomfort? For a profession whose self-definitions and codes of ethics have always stressed the "welfare" of humankind, a missile, no matter how "smart," cannot be a badge of honor.

It helps somewhat to recognize that most engineers are engaged not in weaponry but rather in such life-enhancing works as hospitals, bridges, and concert halls. Yet the stark fact is that until recently 20 percent of American engineers were directly engaged in military projects, and many others were tangentially involved through their work in electronics, computers, and other specialties. Historically the profession's roots are intertwined with military enterprise. The very word, *engineer,* was first used to describe the designers of "ingenious devices," most particularly catapults used by Roman armies. There is some comfort, but not much, in observing that advances in weaponry have made it possible to reduce mass destruction by more accurately attacking military targets. As Daniel Koshland noted in an editorial in *Science* magazine, the concept of "humanizing war" is the ultimate oxymoron.[8]

I've heard it said that technology, through the medium of television, has served to turn Americans away from war by bringing battle scenes into our living rooms. On the other hand, by filtering out the sound, smell, pain, and fear of war, television presents us with theater safely contained behind glass. Among the people I know, the televised version of the Iraq war seemed to evoke horrified fascination, which is something quite different from horror. Still, we can thank technology for making it increasingly difficult to hide war from public view.

I suggest that there is another way in which engineers might find satisfaction in the recent conflict, beyond the obvious but uncomfortably ambiguous pride in technical skill. Operation Desert Storm differed from previous wars by revealing in our society an increased concern for individual human lives. How striking it was that President Bush, General Schwarzkopf, and other political and military leaders shied away from talk of killing. There was not only an unprecedented effort to reduce casualties, but also a determination to avoid discussing them. Perhaps for the first time in the history of war, nobody wanted to talk about body counts. This represents an amazing step forward on the road to civilization, and I would argue that engineers can take a share of the credit. In a world where we have learned to grow food for multitudes, combat diseases, and otherwise save people from untimely death, lives suddenly seem more precious than they did only a generation ago.

As for the Nintendo syndrome, I believe that the concealment of enemy casualty estimates, added to the dazzling high-tech weapons, made it possible to think in terms of computer games. General Schwarzkopf's admonition to the press was not really fair. He had helped create the illusion he disparaged.

And illusion it was. Everybody knows that there was death and suffering aplenty. But that is the point. The pronouncements of our leaders were carefully crafted to reflect evolving concerns of the citizenry. It seems that the people wanted war but not bloodshed. In retrospect, the trench warfare of World War I and the carpet bombing of World War II seemed unspeakably barbaric. The possible use of chemical or biological weapons was viewed with dread and abhorrence. Nuclear weapons were not mentioned.

I've associated this emerging aversion to mass slaughter with the development of life-saving technologies. Perhaps another phenomenon is also at work. During Britain's fifteenth-century War of the Roses, the marauding armies, for all their ferocity, did not engage in civilian massacres as was typical of the age, one explanation being that the war took place on a relatively small island where people, over time, would have to live with each other. As our world becomes "smaller"—another consequence of technological advance—a similar feeling of mutual dependence may be evolving on a global scale. This might move us to devote our technological wizardry—our satellites, lasers, computers, and the rest—to disarmament and monitoring in the cause of peace. The term "Nintendo war" reveals not only our failings but also our aspirations.

## BRIDGES TO PEACE

One of the pleasures of being an engineer lies in the hope that technology, through the prosperity it can produce, will enhance prospects for world peace. Material abundance for all: Might this eliminate the main motivation for neighbor to war upon neighbor? If so, the search for cheap energy, plentiful food, and affordable consumer goods becomes in effect a quest for the Holy Grail.

However, when one considers the fighting that has recently taken place in Bosnia, such a notion seems naive. In that mountainous realm, once a part of Yugoslavia, the slaughter and suffering have been horrifying beyond belief. In the heart of Europe—"civilized" Europe—ancient enemies, ethnic and religious, play out a primal tragedy that seems destined to go on forever. If communal hatred inevitably leads to war, and if such hatred is rooted in history and human nature, then it is illusory to think that technology can help. Of course, order can be imposed by well-equipped police forces, but that sort of solution gives little reassurance for the future of humanity. Small wonder that the latest Balkan conflict, added to long-standing strife in Africa, the Middle East, and elsewhere, gives rise to pessimism. Engineering progress seems powerless in the face of age-old tribal hostilities.

Yet, perhaps we are drawing unwarranted conclusions about cause and effect. One can find many places in the world where ancient animosities burn fiercely yet war is unthinkable. Take Belgium for example. A few years ago I visited there and was astonished to find how bitter are the feelings between many Flemish-speaking citizens and their French-speaking countrymen. This acrimony has occasionally boiled over into demonstration, or even riot, but solutions are inevitably found in political accommodation. The same can be said of Canada. I have spent time with engineers in Montreal, and also in Toronto, and could scarcely believe the hostility that exists between people—even co-professionals—because of language and cultural differences. Yet it would be absurd to suggest that civil war is likely to break out in Belgium or Canada.

What is there about Belgium and Canada that makes them different from Bosnia? Many things, and I don't want to be simplistic; but in the context of penchant for civil war I would say that the biggest difference is the level of prosperity. Bosnia is relatively poor, while Belgium and Canada are relatively rich. Where people are well-fed, reasonably comfortable, and reasonably secure, and where people feel that they have an investment in the community that provides for them and their families, civil war is not a likely prospect. Belgians and Canadians can hate and be prejudiced just like Bosnians. One need not argue that prosperity brings virtue and amicability, merely that it inclines people toward keeping the peace. Even in Northern Ireland, long an archetype of tribal hatred, evidence is growing to support this hypothesis. In late 1994, a *New York Times* headline read: "In Belfast, Prosperity Eases Catholic Nationalism."[9]

Reflecting upon American experience in Somalia, former President Jimmy Carter drew this conclusion: "Civil wars usually develop when neighbors contend for dwindling supplies of food, water, arable land, or a modicum of human dignity."[10] Ethnic hostility is not in itself destined perpetually to beget war.

Beyond poverty, there is another reason why Bosnia is a quintessential candidate for strife. It is mountainous, long isolated from the outside world, and topographically suited to tribal conflict. Slobodan Selevic, a prominent Serbian writer who opposed the aggressive policies of the Serbian president, Slobodan Milosevic, noted that the Serbs who live in the flatlands are not generally belligerent. The mountain Serbs, on the other hand, for centuries were hajduks, or outlaws, resisting the ruling Turkish Empire, and they represent "a different culture."[11]

How does one alleviate the isolation, and possibly lessen the suspicious pugnacity, of mountain people? By building bridges, of course. Ivo Andric, a Serb and a Bosnian who was awarded the Nobel Prize for Literature in 1961, reflects on this theme in *The Bridge on the Drina,* a historical novel set in Visegrad, a village near Sarajevo. According to an ancient myth recounted by Andric, when the earth was first created, its surface was smooth. But the devil, envious of God's gift to humankind, scratched the earth with his nails. Thus "deep rivers and ravines were formed which divided one district from another and kept men apart, preventing them from traveling on that earth that God had given them as a garden for their food and support." To ease the torment that ensued, God sent down a band of angels. They spread their wings across the ravines, enabling people to reach their places of work and to once again communicate with their fellows. "So," concludes Andric, "men learned from the angels of God how to build bridges, and therefore, after fountains, the greatest blessing is to build a bridge and the greatest sin to interfere with it."[12]

Material abundance and the building of bridges: two reasons to hope that technology can be a force for peace. In the midst of disheartening conflict, we should cling to the vision of a world without war, and the engineer's role in making it a reality.

## RESISTING BIG BROTHER

Even if there were an end to armed conflict, there would remain threats to freedom and human dignity. It has long been feared that technology might put sinister social controls into the hands of tyrants, and that that would be the worst fate of all.

This worry has some factual basis, and its fictional portrayal by George Orwell and others has helped to deepen public concern. Yet events of recent years have demonstrated exactly the opposite of what has been feared. Technology has provided an unceasing flow of new weapons for freedom.

I recall how encouraged I felt in 1988, during the time of the Noriega dictatorship in Panama, when I read of expatriates living in Washington, D.C., who sent news reports to their homeland via telefax. To bypass Panamanian government censors, these independence-minded citizens faxed newspaper and magazine articles to several business offices in Panama City, where they were transmitted to churches, schools, labor union headquarters, and other such places for copying and widespread distribution. As expressed by one of the leaders of the enterprise, this process represents "death to dictators, because what can they do to stop transmission?"[13]

The fax is merely one of the latest instruments of social and political liberation. For five hundred years the printing press has been a fundamental component of democracy. The electronic revolution of the past thirty years has made communication vastly more widespread and difficult to repress.

Telephone fax and modem lines can be cut, but total quarantine for a nation is scarcely feasible in today's world. And even if this is attempted, new communications gadgets can easily be smuggled across national borders. The downfall of the Shah of Iran has been credited in part to the simple tape recorder. Messages from the Ayatollah Khomeini were taped at his exile headquarters in Paris, smuggled into Iran, copied and recopied, and then distributed by the thousands on cassettes playable on ubiquitous small machines. In the Iran of today, movies banned by the fundamentalist Islamic government are widely disseminated on videocassette tapes.

In the Soviet Union, before the fall of communism, the "alternative press" movement made effective use of desktop publishing techniques. Lev Timofeyev, the editor of *Referendum,* an unofficial opinion magazine, used a Toshiba 1000 personal computer and a Kodak Diconix printer, purchased by friends at a Moscow second-hand store. "Modern technology will spread and come into its own," said Mr. Timofeyev in 1988. "The authorities already are losing control over the spread of information."[14] When Soviet hardliners staged a coup against the Gorbachev regime, their attempts to control the flow of information were frustrated by privately owned electronic equipment. Army troops outside the Russian parliament building listened to short wave broadcasts from the BBC and the Voice of America. Cellular telephones were used to circumvent controls, as were more than two hundred computer bulletin boards, many linked to the West.

A 1992 uprising in Thailand featured protesters who called themselves "mobs mua thue"—the mobile phone mobs. The evolving computer Internet is used increasingly to bypass government constraints. By means of the Internet, the Digital Freedom Net circulates the work of authors outlawed in their home countries: a Chinese dissident, an Indonesian novelist, and a banned Iranian writer. PeaceNet and other nonprofit on-line services are used by Amnesty International and similar organizations to post notices regarding human rights.

In early 1995, the Chinese government announced that it would create a nationwide computer network linking more than one hundred college campuses to the Internet, even though in the past students at these same campuses have been the center of political dissent. The Communist government of Vietnam is allowing Internet servers to open for business, despite considerable trepidation. "Vietnam has been totally isolated," says a leading computer scientist at that nation's Institute of Information Technology, "and the Internet is the fastest, cheapest way to reintegrate Vietnam into the world."[15]



In India, millions of people are hooking into cable TV systems connected to satellite dishes. In China, small, individual satellite dishes are being installed by the hundreds of thousands. (Government leaders have issued regulations prohibiting their use, but the Ministry of Electronics operates a factory that turns out more than sixty thousand per year.)

Many Americans assume that basic freedoms were won in this country long ago and are adequately assured to all. Yet here, too, new technologies are beneficial. There is evidence that power in the media has become somewhat concentrated, a disconcerting development despite the apparent benevolence of the custodians of such power. The proliferation of cable television channels serves to reverse this trend. Electronic networking, and publishing through computer modems, will help further disperse the power of the press.

The propaganda tools available to authoritarian leaders are formidable. But with advances in electronic engineering, increasing economies in mass production, and almost inconceivable feats of miniaturization, the battle for freedom of information is being won everywhere. Technology, long feared as an oppressor, is proving to be a champion of liberty.

This victory for the human spirit, even after it is achieved worldwide, does not mean the coming of a golden age, but rather leaves us with many difficulties. First, vigilance will be required to make sure that the systems we put in place are maintained and protected. It will be small comfort having a global information network if it hinges on a few satellites that are subject to control by powers that are not themselves subject to control. And we dare not let the manufacture and distribution of crucial elements—microchips, for example—be dominated by any particular group. Security, redundancy, maintenance, and continued availability of information technology to all—these must be our concerns even in the era of the world community.

Then there are the new threats to freedom inherent in increasing technological sophistication: electronic eavesdropping, computer analysis of individual life habits, aerial photo analysis, and so forth. The dangers of Big Brother do not disappear just because the hazards of thought control are found to have been overstated.

Many industrialized nations have adopted legal codes to protect their citizens from excessive invasions of privacy. In the United States we have the U.S. Privacy Act of 1974, which proscribes the government from taking computer data gathered for one purpose and using it for another. For example, there should be no cross-checking on individuals between the Census and the IRS. Also, special numerical identification is not to be established for the purpose of accumulating extensive personal dossiers. Civil liberties defenders are particularly concerned that the Social Security number not be used beyond its original intent.

As for the dangers *to* Big Brother—or to legitimate governments confronted with technically sophisticated criminals and terrorists—they have become ever more daunting. Digitized telephone communication makes wiretapping nearly impossible. The FBI asks for special chips to be used so that an important crime-fighting technique will not be lost. As might be predicted, a heated public debate ensues. Counterfeiting has proliferated with the aid of new engineered apparatus, so countermeasures have to be taken—special papers, inks, and holograms. Plastic and fertilizer-based explosives pose a troublesome challenge in commercial aviation and elsewhere, as we learned at New York's World Trade Center and the Oklahoma City federal building. X-ray scanners adequate for detecting metal weapons, are useless when plastics are involved. Thermal neutron analysis has been tried, but the equipment is expensive, unwieldy, and has had a tendency to give false alarms. Improved X-ray machines are being developed, mechanisms that can detect plastic explosives even within metal containers. Also electronic "sniffing systems" that pick up and rapidly analyze tiny particles from the surface of containers. Such devices are also promising for the interdiction of illegal drugs. But the contest is never-ending, as clever criminals concoct ever more ingenious methods.

Finally, how will we cope with the new capabilities provided us by technology in our daily lives? What will we do with our increasing freedoms? Do we want to use electronic polls that can instantaneously reveal how every citizen feels about any given issue? How much pure democracy can we absorb? Do we want to tolerate automatic telephone-dialing devices that, according to a lobbying group called the Direct Marketing Association, already reach more than 7 million Americans daily? If not, how and by whom are they to be controlled? How about copyright infringement in the age of the magnetic hard drive; telephone caller identification; and security for E-mail?

Our apprehensions have been misplaced. Technology has been depicted as a Frankenstein's monster that, having been created, might control and destroy us. A more apt parable is the genie in the bottle. After being released, the genie neither threatens nor constrains, but rather offers to grant wishes. Technology, like the genie, confronts us not with slavery, but with something nearly as vexatious: freedom of choice.

CHAPTER 5

# AMBIVALENCE AND DOUBT

## COPING WITH MOTHER NATURE

Sometimes the engineer, like any thoughtful person, experiences doubts and apprehension. Global warming, ozone depletion, acid rain, deforestation, and species extinction—these are a few of the topics that must give one pause. Have we gone too far in tinkering with the natural world? Are we being reckless? Should we think about "resacralizing" nature, as Jeremy Rifkin says we must, in an anguished book called *Biosphere Politics?* From time to time, my engineer's zest for progress is tempered by profound concern for the natural environment.

But, just as I'm about to give way to qualms and indecision, along comes some terrible natural disaster and my mood changes. Take, for example, the cyclone that attacked Bangladesh in 1991, killing 125,000 people and leaving some 9 million homeless. That catastrophe aroused in me feelings of anger and betrayal as well as dismay. How can one cultivate a more compassionate attitude toward "the environment," when confronted with such a violent reality?

Scarcely had the photos of corpses and suffering survivors of the cyclone vanished from the newspapers when a volcanic eruption in Japan became a front-page story. A flood of superheated ash and toxic gases from Mt. Unzen exterminated a farming village called Kita-Kamikoba, killing about fifty people, and threatening 45,000 inhabitants of the nearby city of Shimabara. In 1792, Shimabara had been swept into the sea when a volcanic eruption was followed by a tidal wave, and the current residents fear similar cataclysm. Events concerning the volcano were followed intently on television by millions of Japanese. For all their nation's technological and economic strength, many Japanese hold the view that they live on a fragile island, ever on the verge of disaster, and the travail at Mt. Unzen evoked a fervent communal response. As if to emphasize the precarious nature of life on earth, another volcano—this one in the Philippines—erupted the following month. More recently the Japanese have been struck by earthquake in Kobe, and closer to home, there have been major floods in California and along the Mississippi River.

Truly, we all live on a fragile island ever on the verge of disaster, with storms, volcanoes, earthquakes, and floods that have throughout history leveled cities and killed millions. Perhaps even worse than such periodic catastrophes are the unremitting afflictions of drought, famine, and plague. The 1990s have brought us these in full measure.

In sub-Saharan Africa, 29 million people face starvation in what Save the Children, a relief agency, calls "the worst famine in Africa in living memory."[1] A woman in the Sudan explains, "There was no rain this year, so nothing grew."[2] It is as brutally simple as that.

In Latin America, a cholera epidemic is out of control, threatening 6 million people, forty thousand of whom might die. And in New York City, where I live—as if evidence were needed to show that no island is immune from calamity—the scourge of AIDS continues in its second decade.

Given this litany of woe, I can only conclude that nature, if not actually malevolent, is certainly pitiless. My benign sentiments seem misplaced. Is not nature an enemy? Must not human beings—engineers first of all—fight against this enemy for the sake of survival? Perhaps the seventeenth-century English philosopher, John Locke, had a point when he said, "The negation of nature is the way to happiness."

Clearly we are not able to divert cyclones or protect the lowlands of Bangladesh from flood tides (unless we build hundreds of miles of dikes at a cost beyond imagining.) But we *have* made spectacular advances in forecasting the weather, using satellites and computers, and in communicating our findings, using radio, television, and portable loudspeakers. Lives can be saved through evacuation plans tied to early warning, also by construction of raised, storm-proof sanctuaries. As for rescue operations, the airplane—especially the helicopter—is a man-made angel of mercy, and its deployment, along with food and medical supplies, should be central to international disaster planning. The same sort of thinking applies to earthquakes and volcanic eruptions. A robot developed at the Robotics Institute of Carnegie Mellon University, and sponsored by NASA, is being used to investigate the interior of active volcanoes. By sampling the venting gases, geologists and seismologists hope to improve prediction of eruptions, and also study how these gases might contribute to ozone loss and global warming. Earthquake-resistant structural design is a well-advanced specialty that may show the way toward coping with seemingly irresistable natural forces.

Even famine can be made to yield to technology. There *is* enough food in the world, and agronomists are developing ever hardier plants, including strains that can thrive with very little water. Eventually, we will turn saltwater into fresh economically enough to enable us to grow crops almost anywhere. At the moment, the problem is distribution, sending food from where it is plentiful to where it is scarce. Here the difficulties are political as well as technical, and as events in Somalia and elsewhere have demonstrated, solutions can be devilishly difficult to find and apply. Nature is cruel to withhold the rain, but we are equally cruel to withhold sustenance from our fellow humans.

As for disease, what can one say other than we need research and more research, along with advances in research equipment and techniques. The war against nature's devastations must be unremitting.

Of course, even as we utter it we know that war is not the appropriate word. If we use the blind hostility of nature as a rationale for all-out combat, including debasement of the environment, we will only be adding to our perils. On the other hand, we dare not let poets and environmentalists lull us into acceptance of our fate. If humanity is to endure, we must resist the onslaughts of nature with all our vigor and ingenuity.

If enemy is too harsh a concept, the term Mother Nature is surely too tender. Pliny the Elder wrote in his *Natural History,* almost two millennia ago, "It is far from easy to determine whether she has proved to man a kind parent or a merciless stepmother." Or, as Norman Maclean, an ardent naturalist, once wrote, quoting a science teacher with grudging agreement: "The universe, she is a bitch."[3] Perhaps it would be just as well to stay away from the metaphor of womanhood altogether.

Some of my favorite philosophizing is to be found in the comic strip, "Calvin and Hobbes"; and one of my favorite episodes finds young Calvin outside in the middle of a rainstorm. His plans for the day having been ruined, Calvin shouts angrily at the sky: "Do your worst! C'mon, let's see what you've got! You can't crush the human spirit! On behalf of all earthly life, I defy you!!" To emphasize his temerity, Calvin takes off his clothes and splashes in the puddles. But he hadn't reckoned on the hailstorm that suddenly arrives. "Ow!" he yells, "Are you trying to kill me?! Ow! What's wrong with you?!" As he races into his house, his long-suffering mother stands by the front door. "I'll bet there's an explanation for this," she says, "and I'll bet I don't want to hear it." Calvin explains: "The universe has an attitude, Mom!"[4] It often seems that way.

To go from the ridiculous to the sublime: In the Book of Genesis it is told that Jacob wrestled with an angel and would not let him go until he secured the Lord's blessing. Wrestling is neither war nor submission, nor does it remind one of children or parents. I suggest that the metaphor is apt for describing the relationship between engineers and the natural world.

## COSMIC HUBRIS

Consulting the Bible inevitably leads one to ponder where, in the very long run, we may be heading. Does engineering have anything to say about *Truth* with a capital *T?* And if so, how does that affect society's view of the profession and its works?

When, in this modern era, philosophers turn to the ultimate questions, it is science rather than technology that they examine. Yet the phrase "science and technology" has become such a standard term that technology, by association, is swept into the discussion.

I have made many claims on behalf of engineering, and I will make more in the following pages; but I do not believe that engineering provides answers to the mysteries of the universe. There are many wonderful things that both science and technology bring to our lives, but answers to the ultimate questions are not among them. Unfortunately, in trying to "sell" science to the public, hucksters have made promises that can never be kept. This is a shame, since there are so many legitimate ways to excite public interest. Through excessive zeal and deceptive marketing the cause of science has been hurt. The cause of engineering has been tainted as well.

Consider a mid-1990 issue of *Time* magazine. "SMASH!" declares the cover in large letters on a background denoting explosive atoms. In smaller but still conspicuous print there appears a pulse-quickening announcement: "Colossal colliders are unlocking the secrets of the universe."

The secrets of the universe! How can any thoughtful person resist picking up the magazine and plunging right in? There the article is headlined, in the same feverish tone, "The Ultimate Quest."[5]

The text, however, will disappoint anyone in search of prophetic secrets. It is a straightforward report on the scientists who are trying to discover the subatomic particles thought to be the basic components of matter. At the Fermilab Tevatron in Illinois, the Linear Collider at Stanford, and CERN's Electron-Positron Collider outside Geneva, researchers smash streams of particles against each other in search of a comprehensive physical explanation of the stuff of the universe. If the Superconducting Supercollider had been built as planned in Texas, physicists had hoped to duplicate conditions that existed in the earliest fractions of a second after the Big Bang, in effect to re-create the state of matter at the beginning of the universe.

This is very exciting science, but hardly grounds for the quasi-religious rapture that it seems to arouse in publishers and headline writers.

The Hubble Space Telescope, launched in 1990, evoked similar editorial excitement. An article about the marvelous contraption in the *New York Times Magazine* was entitled, "The Search for the Beginning of Time."[6] A sub-caption promised that "a point of mind-stretching revelation is at hand." *Revelation,* a word that is steeped in transcendental implication, becomes a watchword for our time.

Wherever scientists are searching for the fundamental architecture of the natural world—in micro realms or macro—they are accompanied by camp followers who attempt to define scientific work in frenzied terms of spirituality. An advertisement for a serious book by an astrophysicist about "dark matter," promises access to "THE DARK HEART OF THE UNIVERSE." A dinner program for alumni at a major university, prepared by the chairman of the astronomy department, is publicized as "THE MYSTICAL UNIVERSE." A philosopher writes a book called *The Quantum Self,* with a startling subtitle: "Human Nature and Consciousness Defined by the New Physics." Similar works include *Lonely Hearts of the Cosmos: The Scientific Quest for the Secret of the Universe, The God Particle,* and *Theories of Everything: The Quest for Ultimate Explanation.* A review of such books is headlined on the cover of the *New York Review of Books:* "The Big Secret of the Universe."[7]

Nor is this fervor restricted to physics and astrophysics. Chemists and biologists, delving ever deeper into the world of DNA, are heralded as approaching elemental knowledge of the nature of life. Artificial Intelligence, as discussed in the public press, verges on the supernatural. Magazine articles about geology, paleontology, and archeology stress "origins" and "beginnings," as if secrets of the universe were to be found at the bottom of an excavated hole.

Science at its best has always been exciting. But its essence has always been dispassionate analysis. And scientists have found that each discovery leads, not to a blinding epiphany, but to the unfolding of new mysteries. It is ironic that even as leading scientists, and the best science writers, stress the infinite nature of the enterprise—"our knowledge of our ignorance," to quote Karl Popper—so many who interpret science for the public are talking in terms of definitive truth. This fervor goes beyond the utopian literature of earlier times: The message now is not merely that science and technology will ease our lot and improve the human condition. We are hearing promises about ultimate verities.

Yet isn't this harmless enough, one might ask, since the hard sell and puffery have long been features of our commercial society? But such a light-hearted view overlooks the harm that might be done to science if too much is expected of it. Anyone who has been beguiled by false promises may, like a jilted lover, succumb to resentment, and even anger. The well-being of science depends in large measure on public support; it is clear that harm can be done by fervent promoters. Although most people do not engage in metaphysical discourse, many do yearn for ultimate truths. We should not assume that those who casually exploit this very human trait will escape retribution.

Albert Camus once told of the excitement with which he learned about the theory of the atom, in which electrons were depicted gravitating around a nucleus. But then came the inevitable letdown: "You explain this world to me with an image. . . . What need had I of so many efforts? The soft lines of these hills and the hand of evening on this troubled heart teach me much more. . . . I realize that if through science I can seize phenomena and enumerate them, I cannot, for all that, apprehend the world."[8] It has been more than half a century since Camus wrote those words, and there have been many advances in scientific theory, but none that would change in the slightest this basic insight, or the melancholy alienation from science it betokened.

The human heart does indeed crave to "apprehend the world." But science cannot yield the ultimate "knowing" that is sought. Science, for all its marvels, cannot take the place of art, religion, and philosophy. Overzealous boosters who do not recognize this do harm to the enterprise they claim to serve.

## LOST IN (CYBER)SPACE

Just as I am distressed by the misguided search for God in science, so am I troubled by the growing fervor that attends the evolution of computer networking.

I'm excited about the prospect of the information highway. I'm thrilled by the thought of being able to search the world's libraries while sitting at my desk at home. I'm delighted with the possibility of ordering up the movie of my choice, the music I'm in the mood for, or the sights and sounds of faraway places. Although I don't care for E-mail, electronic bulletin boards, or networking with thousands of strangers, on the whole these developments seem benign, perhaps even good for democracy. And, of course, as an engineer, I'm proud of how my profession has brought such marvels to a world much in need of practical help—to say nothing of a psychic lift.

Yet along with these pleasant thoughts I feel apprehension. This stems from the feverish descriptions I've seen of a mystical realm called "cyberspace." Apparently one actually "enters" this vast cosmos, moving through computer screen windows, and then "explores," following one's fancy. Call up the text of *Moby Dick,* then summon the sight and sound of the ocean, research the life of Herman Melville or the anatomy of whales, see Gregory Peck in the movie and freeze a few frames for later use—this is the sort of thing that awaits us in the new enchanted galaxy. According to Steven Levy, author of a book about the Macintosh computer, millions of people already "make the excursion" daily. In the future, he predicts, "we will cross the line between substance and cyberspace with increasing regularity."[9]

I like new metaphors, and the idea of following one's every impulse or fleeting idea, wherever it may lead, has positive implications for nourishing creativity. But as I contemplate this new form of intellectual space travel, I am struck with a case of vertigo. I fear that we may be undercutting the mental processes upon which good engineering depends. Caprice and inspiration are important, but so is the ability to assess experimental evidence, comprehend mathematical verities, and refute false appeals to intuition. Technical work requires discipline, concentration, and restraint. We should be cautious about losing these traits among the whirling universes of infinite choice.

Few of the people I know share my uneasiness, least of all the engineers and educators who should be most concerned. Optimistic delight seems to be the prevailing sentiment. In the words of a professor of learning research at MIT, mere children will steer a "knowledge machine" to any topic of interest, "quickly navigating through a knowledge space much broader than the contents of any printed encyclopedia."[10] It will be fun, I do not doubt, sitting in front of workstations and browsing through almost limitless hypermedia databases. And as for students, the promises of "self-guided learning" and "knowledge by exploration" sound wonderful.

Yet we cannot constructively spend much of our time under the spell of serendipity or free association or even purposely flitting from one interest to another. The most useful thing I remember learning in school was how to prepare an outline; how to organize my thoughts; how to *focus*. The amount of available information is incredibly vast, the day is short, and we are, after all, human. An average educated person can read only about 360 words a minute, speak 150, hear and comprehend 250, and type 60.

In his book, *Uncommon Sense: The Heretical Nature of Science,* physicist Alan Cromer argues that the formal thinking needed for math and science is not physiologically normal in humans but needs to be inculcated by education. He maintains that scientific thinking, which is analytic and objective, "goes against the grain of traditional human thinking, which is associative and subjective." Citing the work of the Swiss psychologist Jean Piaget on the mental development of children, he notes that infantile egocentrism, confusion of the self with the outside world, is the major obstacle to mental growth. The ability to understand mathematics and science is especially compromised by failure to overcome "the innate human tendency to confuse thought and reality."[11] I fear that flights through cyberspace, however energizing they may be for the imagination, may weaken the objective rationality needed to do good engineering. The danger is greatest for children—the potential engineers of tomorrow; but none of us is safe.

I don't know what we can do to guard against the allure of cyberspace except to caution others—and ourselves—about its dangers. Perhaps we might start by floating an icon on our computer screens: Odysseus tied to the mast so that he could hear the song of the sirens but not be diverted from his journey.

## THE SPECTER OF OVERPOPULATION

Somehow we learn to cope—with unwarranted criticism, unsettling ambivalence, and annoying people who oversell technology or misuse it. But there are two problems that will not yield to rational self-assurance—two threats looming like dark clouds that will not clear. The most ominous of these is overpopulation.

As the reader will have recognized by now, I am, like many engineers, optimistic in the face of technological challenges. No matter how daunting the technical obstacle, I instinctively presume that human resourcefulness will be equal to it. A lifetime of experience has shown me that engineering solutions are likely to appear on the scene when needed—often imperfect and carrying within them the seeds of new difficulties—but solutions nevertheless.

Each day brings new reassurances. Is energy a problem? There is word of an exciting new breakthrough in photovoltaic cell design at the University of New South Wales in Australia. The new approach features very thin silicon layers that can convert solar energy to electricity at relatively high efficiency using low-cost impure silicon. This promises to cut costs by 80 percent, making solar technology competitive with conventional power sources. Progress is also reported on the frontiers of fusion, and surely, before long, we can expect new breakthroughs in superconductivity. With energy that is affordable and environmentally benign, many formidable problems—including the need for warm homes and potable water—become less worrisome.

But if world population doubles again in the next forty years, as it did in the last forty, how can our energy resources possibly be adequate to the need?

As for food, a report from the Council for Agricultural Science and Technology predicts that with "smarter farming," crop yields can be vastly increased while at the same time allowing much land to be returned to the wild. In India, the adoption of high-yield grains has already enabled farmers to "spare for nature" more than 100 million acres that otherwise would have been plowed and planted.[12] The prospects for increasing food output through technology are heartening. But not if we find ourselves with twice the number of mouths to feed.

The environment? A book issued by the National Academy of Engineering, called *The Greening of Industrial Ecosystems,* contains essays on such topics as industrial ecology, wastes as raw materials, design for the environment, and preventing pollution. As ecological consciousness is joined to engineering genius, great improvements will be achieved. We can save the environment, particularly as we devise benign technologies that generate plentiful energy, water, and food. It takes confidence verging on hubris to think this way, but the "can do" attitude is deeply ingrained in the engineering psyche, and rightly so. Our successes are many.

Nevertheless, confronted with the prospect of exponentially increasing numbers of humans, my optimism fails.

The world population, now 5.7 billion, is double that of the late 1950s. Experts fear that the figure could grow to 12.5 billion by the middle of the next century. There are some people, engineers included, who believe that we can cope with even that drastic growth, but I am not one of them. I am convinced by United Nations demographers, and other authorities, who predict that famine, wars, and unspeakable suffering will overtake the world if population is allowed to grow unchecked. I also feel this through intuition, which, for all its potential flaws, is a time-honored element of engineering judgment.

The third International Conference on Population and Development (ICPD 94), sponsored by the United Nations, was held in Cairo in September 1994. The goal established by conference organizers is to follow a low-growth path to a world population of 7.27 billion in 2015 and finally to achieve stabilization at 7.8 billion by the year 2050. From 5.7 billion to 7.8 billion—an increase of 37 percent over a period of 55 years—is a challenging prospect, but at least it can be realistically addressed.

These figures, although doubtless optimistic, do not come out of thin air. They are based upon past successes and actual possibilities. The conferees at ICPD 94 did not start from square one, but rather hoped to build upon national population policies and family planning programs developed over the past twenty-five years. In most industrially advanced parts of the world, fertility rates are already below the "replacement" level of two children per woman. In developing nations, birth rates have been lowered from over six children per family in the 1960s to about 3 1/2 children today. Approximately half this decline is attributed to contraceptive use, which has increased five-fold in developing nations over the past several decades and now stands at 55 percent of married couples.

In addition to traditional family planning approaches, experts stress the importance of many interrelated issues: sustainable economic growth, universal access to health services, reduction of infant and child mortality rates (lessening the perceived need for large families), education and job training, and—a concern of growing force—gender equality and the empowerment of women. Commitment is essential, both from individual governments and from the world community. Money, needless to say, must also be forthcoming.

Engineers had good reason to follow the proceedings of the Cairo conference, and in its aftermath to work for its objectives. Although there are medical and technological advances yet to be made, it is clear that many of the most immediate challenges are political and social. Accordingly, it is essential that members of the profession speak out strongly on the issue. If world leaders can be made to see the future as engineers see it, efforts on behalf of population control will not be wanting. The challenges posed by population growth will be enormous. We don't want them to become insurmountable.

## SOCIAL LIMITS TO GROWTH

The other dilemma that troubles me, even as I look at the world through the rose-tinted glasses of technological optimism, is the open-ended nature of human ambition. Suppose that, through engineering, we eventually provide for the basic needs of all the people on earth—and more besides for leisure time, amusement, culture, and education. Human cravings will still remain unsatisfied. In fact, each time we achieve a success, engineers expose new possibilities and beget new appetites. I do not refer to ineffable aspirations, such as religious insight or eternal life, which cannot be engineered. I refer to everyday physical desires that engineering can help to satisfy—for some people, but not for all.

There are absolute limitations on *social* consumption that come into play long before we encounter problems in *physical* consumption. Consider this example from Fred Hirsch's thought-provoking book, *Social Limits to Growth:*

> An acre of land used for the satiation of hunger can, in principle, be expanded two-, ten-, or a thousand-fold by technological advances. These advances may occur in one or all of the processes that come between the productive agricultural use of that acre and the end product in the form of nutrient. From the same productive acre, more and more food can be and has been produced. By contrast, an acre of land used as a pleasure garden for the enjoyment of a single family can never rise above its initial productivity in that use. The family may be induced or forced to take its pleasures in another way—substitution in consumption—but to get an acre of private seclusion, an acre will always be needed.[13]

Even today, the desire for private seclusion is increasingly difficult to satisfy. Indeed, limitations in this area will soon be upon us, if they have not been reached already. Happily, the taste for seclusion has not become universal.

Most troublesome are those yearnings the essence of which is to have something that other people do *not* have. We might note with satisfaction that silk stockings, which were once the privilege of queens, are now available to everyone. The same can be said of books and access to lavish entertainment. True, but the nonmaterial aspects of privilege cannot be shared in the same way. There is nothing technology can do to help everyone feel superior to his or her neighbors, to have servants, or to have the single best seat in the arena. It is only in Garrison Keillor's Lake Wobegon that "all the children are above average."

But what, one might ask, does any of this have to do with engineering? Personal ambitions, and the conflicts they engender, are as old as recorded history. Yes, but in earlier ages most people were willing to "know their place" in the social order. Through the marvels of technology we have contributed to the rising expectation that each person is deserving of a choice place at life's banquet table.

Further, in the United States we have developed an individualistic ethos, assuming that the pursuit of self-interest contributes to the common good. Engineering has worked hand in hand with capitalism, and the marvelous results are manifest. Competition is given a place of honor in our national pantheon, as well it should be, considering the good effect it has had. While individualists have been trying to build better mousetraps under the banner of free enterprise, the standard of living has risen for all.

But competition carried to its ultimate extreme—fierce competition for overweening privilege—becomes, in Hirsch's terms, "excess competition." If this develops beyond acceptable bounds, "individual freedom will then be seen to be socially destructive and ultimately self-destructive, and pressure to restrict such freedom will become irresistible."[14]

The competition need not be aggressive to be repugnant. We are all familiar with tales of business pioneers whose grandchildren become self-absorbed consumers, spoiled by the very success achieved by their forebears. Technology overindulges the many the way a wealthy family overindulges a few.

Although engineering, by extending general welfare, has helped to loose the forces of egoistic ambition—and has brought our society closer to vexing social limits—I do not suggest that we seek engineering solutions to the problem. That is, I do not favor social engineering, or as it is sometimes called, *technocracy*. Let us hope that democratic government, plus common sense, will keep us one step ahead of social turmoil, as it has until now.

Speaking of common sense, the engineer's way of looking at the world—the "no free lunch" perspective—is a great help in approaching all problems, social as well as technical. Also, engineering, in its own essence, contains elements of a solution to the social limits dilemma. Engineering work compensates the individual who undertakes it with satisfactions that do not threaten the social order. Engineers as a group are not highly paid, nor do they work primarily for money or fame. Some of them are competitive, but mainly in ways that most people (in the Western world, at least) would deem wholesome. Many of them work, studies show, because what they do is intrinsically interesting, and because they enjoy making a contribution to the common good. This attitude—which is shared by teachers, nurses, forest rangers, musicians, social workers, and many other groups—is the most reassuring response I know to the danger of social limits.

Another check and balance that comes into play is the phenomenon sometimes called "reverse snobbism." When arrogance and vanity reach outlandish proportions, they are liable to be ridiculed and scorned. Blue jeans become more chic than fancy duds. Engineers are often found among those to whom ostentation is unappealing. (More about engineers as people in Part II ahead.)

In Asia there is a tradition of suppressing individual ambition in favor of group harmony. This has appealing elements, but in several important respects it runs counter to the American spirit. Is there not a disposition more compatible with our own to which we can look for guidance and encouragement?

During the winter Olympic Games of 1994 there came a time when the host country, Norway, had won so many medals that national pride became tempered with good-natured embarrassment. It was then that I heard commentators speak of a Norwegian ethic that rejected the idea of anyone having too much more of a good thing than his neighbor. It isn't "seemly," it isn't "right," to be perceived as being overly ambitious.

I have long been fascinated by the Scandinavian countries, and thought of them as a potential model for the world's technological future. They are relatively wealthy and highly committed to technology. They have high standards of living. Yet their citizens appear to respect nature, love the outdoors, and passionately support environmentalism. The Norwegians, the Swedes, the Danes, the Finns—they seem to be universally respected as good world citizens, democratic, law-abiding, committed to peace and worthy global causes. Yet, their ancestors, the Vikings, were ruthless marauders. And well into the last century these nations were involved in wars like all the other Europeans. Yet here they are, as technologically sophisticated as any societies on earth, yet apparently unspoiled by the coming of modern machinery, presenting something of a paragon. They give the lie to naysayers who equate technological progress with the decline of decency and agreeable community. They give engineering a good name.

To be sure, Scandinavia presents a special case: homogeneous populations, geographical isolation, special natural resources, and other unique blessings of nature and history. Nor are Scandinavian nations without problems, many of them linked to the coming of the modern age. I have seen youths in Stockholm as dissolute as youths anywhere. And, while it may not be true that the suicide rate in Sweden is the highest in the world, as periodically reported in the press, prosperity and clean streets plainly do not eliminate human unhappiness and clinical depression. Also, the Scandinavian politeness that is so appealing often goes hand in hand with a stuffy self-righteousness. The oppressive social atmosphere that one finds in an Ibsen play persists in many parts of these wonderful lands.

Nevertheless, there are enough encouraging portents in Scandinavia and elsewhere to give one hope.

We must find the spirit and the intelligence to cope, first, with the clear danger of world overpopulation, and next, with the more subtle but still worrisome social limitations to growth.

Whenever I get to thinking of these vexatious problems, and the challenges they pose for the technological optimist, I am reminded of a story that is told about Margaret Fuller and Thomas Carlyle. Margaret Fuller was a mid-nineteenth-century American writer and intellectual, well known in her time, a founder, along with Emerson and Thoreau, of the Transcendentalist movement. One day, according to the story, Ms. Fuller startled a group of her friends by announcing passionately: "I accept the universe!" Upon being told of this pronouncement, the great British historian, Thomas Carlyle, is said to have answered: "By God! She'd better."

Yes, we had all better accept the universe, with all its perplexities and threats as well as its splendid possibilities. As every engineer knows, the first step toward the solution of a problem is a clear and realistic recognition of its parameters.

CHAPTER 6

# IMAGE AND REFLECTION

## THE HOSTILITY OF POETS

I readily admit that engineering cannot solve the ultimate philosophical questions. And I empathize with questing spirits like Albert Camus who are saddened by their inability to find solace in the truths of science and technology. But I am exasperated by those poets and other creative writers who see technology as the *source* of their personal anguish. Is it inherent in the scheme of things that the special people who celebrate blossoms and rainbows, and explore recesses of the human heart, must view the engineer as an adversary? Even when they recognize that engineers, like most people, are disturbed about environmental degradation, and are seeking a cleaner, safer world, their hostility will not subside. Practical assurances do not speak to a cluster of concerns that lie deep in the artistic psyche.

In the *New York Times Book Review* an essayist regrets that romance is not what it used to be. Science and technology, it seems, are largely to blame. We are told that "technology is the knack of so arranging the world that you don't have to experience it."[1] In the *New York Review of Books* a reviewer speaks of engineering as "an alien power, crippling the sense of freedom that it was intended to serve."[2] The editor of an anthology of contemporary poetry states that "The making of community against anti-social technology is the chief object of the poetry gathered here." And further: "It's been our generational lot to sift through the debris of industreality [sic] to force reality through the cracks."[3]

Bill McKibben writes a book called *The End of Nature,* and his elegiac dread of technological change is voiced in the title. McKibben tells of retreating to an Adirondack wilderness where he finds sublime detachment swimming in a lake. But if he hears a motorboat, the mood is broken. When technology intrudes, "you're forced to think, not feel."[4]

One cannot quarrel with a person who values feeling, nor with someone who speaks out on behalf of love or freedom. But why cast technology as the enemy? There have always been obstacles to achieving the euphoric states so celebrated in song and story. People faced with the harsh realities of earlier times—such as plagues, failing crops, or marauding war parties—must often have been distracted from amours and blissful swimming sessions. Technology doubtless has changed the nature of life's diversions—both pleasant and unpleasant—but I do not see why it must be viewed as hostile to artistic fervor.

I do not expect contemporary poets to echo Rudyard Kipling's paeans to machinery, nor to emulate the great yeasayer, Walt Whitman, "Singing the strong light works of engineers." I forgive them if they fail to see the passion inherent in engineering enterprise and fail to incorporate it in their work. And, as I have already said, I understand if, like Camus, they express angst in the face of an unresponsive universe. I merely want them to stop treating technology as if it were the chief source of human discontent.

Little enough to ask, one might think, yet repeatedly I find reasons to feel discouraged. Even as I tap away at my word processor I'm reminded of Wendell Berry's diatribe against this harmless gadget in his book of essays, *What Are People For?* Mr. Berry chooses to continue writing with a pencil, not merely out of personal preference, but as a statement of protest against high technology. His wife transcribes his work on a typewriter bought in 1956 which is "as good now as it was then."[5]

Yet all is not lost. "I use a word processor," declares John Updike, surely one of the most distinguished writers of our time, "and the appearance on the screen of the letter I just tapped seems no more or less miraculous and sinister than its old-fashioned appearance after a similar action, upon a sheet of white paper in my typewriter." Machines simply do not bother Updike the way they do Berry. Having given the matter considerable thought, he states the reason: "The capacity of human beings to absorb what they wish to and to ignore the rest seems to me almost illimitable." He does not believe that the human spirit is threatened by the computer, or by the carburetor that responds "when I simplemindedly ask my automobile to go," or by the telephone that has not perceptibly changed the things that people talk about.[6]

No engineer can speak to the concerns of creative artists as effectively as can one of their own kind, and so I find comfort in responding to Berry, McKibben, and similarly troubled spirits, by referring to the wisdom of John Updike. Perhaps they can cultivate the capacity to ignore those aspects of technology that do not answer to the cries of their heart.

Happily, Updike is not the only poetic creator who has found it possible to cope with a technological world. I have long taken pleasure in a beautiful, life-affirming statement made by the great novelist and Nobel Prize winner, Saul Bellow:

> A million years passed before my soul was let out into the technological world. That world was filled with ultra-intelligent machines, but the soul after all was a soul, and it had waited a million years for its turn and did not intend to be cheated of its birthright by a lot of mere gimmicks. It had come from the far reaches of the universe and it was interested but not overawed by these inventions.[7]

## A HERITAGE OF ANGST

Although it sometimes appears as if the artist's dread of technology is today reaching new heights, there is a time-honored tradition of angst that dates back at least to the story of the Tower of Babel. As the industrial revolution evolved during the nineteenth century, William Blake raged about "satanic mills," while Matthew Arnold lamented that "things are in the saddle." Even in early twentieth-century America, when enthusiasm for the promise of technology was at its peak, a sense of trepidation was finding expression in the arts.

Of all the artistic visions of technology that I have seen or read, the one I've found most haunting is Eugene O'Neill's play, *The Hairy Ape,* written in 1922. Even in the glory years of American engineering—1921 and 1922 saw the introduction of home radio sets and home movies, Technicolor and rayon, escalators and Band-Aids—O'Neill pictured the technological world in apocalyptic terms.

The play's main character is Yank, lead stoker in the boiler room of a luxury ocean liner. As he shovels coal into the roaring furnace he fancies himself a basic life force, the prime industrial mover: "I'm de start! I start somep'n and de woild moves! . . . I'm steam and oil for de engines; I'm de ting in noise dat makes yuh hear it; I'm smoke and express trains and steamers and factory whistles . . . All de rich guys dat tink dey're somep'n, dey ain't nothin'! Dey don't belong. But us guys, we're in de move, we're at de bottom, de whole ting is us!" Not very gracefully expressed, yet a statement that resonates in the heart of the engineer.

O'Neill, however, clearly views Yank and his fellow workers in a different light. These are the stage directions set forth in the script: "The men themselves should resemble those pictures in which the appearance of Neanderthal Man is guessed at. All are hairy-chested, with long arms of tremendous power, and low, receding brows above their small, fierce, resentful eyes."

As the play unfolds, a young female passenger in a white dress visits the boiler room and recoils in horror. Upon seeing Yank, she cries, "Oh, the filthy beast!" and promptly faints. It is a powerful scene when skillfully staged. The lady is effete, a member of an upper class that is clearly ill prepared for the new age. On the other hand, technology, in O'Neill's image, is a subhuman force that rages far beneath us in a hidden boiler room.

Stokers are not engineers; but this vision of the industrial enterprise has far-ranging implications. Society will not be well served if sensitive people come to think of technology as an unpleasant manifestation of primitive energy, something serviceable but gross.

## THE REPROACH OF THE LITTLE TRAMP

While O'Neill was depicting technology on the stage in terms of subhuman bestiality, a very different image was being presented on the screen by that great master of comic mime, Charlie Chaplin. The little tramp, as portrayed in the film, *Modern Times,* is perhaps one of the most scathing critiques of technology ever devised.

I was reminded of the lasting quality of this masterpiece by a group of college students who were studying it intently although they were several decades removed from the era in which it was produced. In their view, Chaplin was an artist of great foresight, who saw the dark side of progress and depicted his wistful Everyman as the slave of a grim mechanical imperative. We are just today beginning to appreciate Chaplin's vision, they argued, as the tyrannical nature of technology becomes clear.

In the film's most famous scene, it is Charlie's job to tighten bolts on a series of steel plates moving steadily in front of him. An ill-timed sneeze causes him to miss one plate as it goes by, and racing after it he dives onto the assembly line belt and is swept into a chute. He then is seen traveling through an immense machine, carried along by cogwheels and rollers. After being freed from the machine he goes berserk, dashing about, turning imaginary bolts on his fellow workmen's noses and attempting to tighten suggestively placed buttons on a woman's dress. He darts into a control room where he creates havoc by randomly flicking switches, and then embarks on a lunatic dance, squirting oil from a nozzled can at everyone in sight. An ambulance arrives; Charlie is hustled into it and driven away.

The scene is virtuoso slapstick, and the satirical critique of factory mechanization is valid and telling, particularly if we recall that the film was made in 1936, when work on many assembly lines was grim indeed. But my young critics insisted on giving the film a deeper, more sinister, meaning, calling the tramp a timeless archetype of humanity confronted with the machine. I found myself on the defensive and obliged to look at the matter with a seriousness to match theirs.

Is the Chaplin character indeed a symbol of the human spirit besieged by a demon technology? If so, it is a sad day for the human spirit. For, viewed through unsmiling eyes, the tramp is seen to be not so much wistful as pathetic. He is not a normal person driven mad by the assembly line; he is unbalanced to begin with. As described by Chaplin in his autobiography, the famous character in baggy pants and derby hat was conceived of as a "tramp, a gentleman, a poet, a dreamer, a lonely fellow. . . ."[8] He has no family, no friends and no home. He is a lost soul long before he sets foot in the infamous factory.

*Modern Times* is remembered mainly for the wild assembly line scene. Few people recall that as the film progresses some of the workers go out on strike, demanding higher wages and better conditions. But not Charlie. No sooner is he recovered from his nervous breakdown and released from the hospital than he encounters a new misadventure. A flag drops off an explosives truck, and Charlie, picking it up, waves to attract the driver's attention. Just at that moment a mob comes marching around the corner, and Charlie is arrested for leading a "red" demonstration. Chaos is his milieu, bewilderment his destiny. He is a clown. There is no world, technological or otherwise, in which he can avoid his bittersweet fate.

I reasoned along these lines with my student friends, but to little avail. Contemplating comic genius it is fruitless—and probably graceless—to argue the facts. It is a measure of Chaplin's art that there is something of the little tramp in each of us, which means that there will always be a touch of trepidation in our attitude toward the machine.

## PRINCE CHARLES AT HARVARD

When philosophers and poets announce their disaffection from technology, I may be frustrated and wish it were otherwise; but in my heart I understand. There are "differences" in the world, and they cannot always be reconciled, much less harmonized, in every individual. But when *academics* disparage technology, I begin to reach the limits of equanimity. And yet it happens all the time, even though engineering has been an exacting academic discipline for well over a century. An irritating reminder of this state of affairs occurred during Harvard University's 350th anniversary celebration in the fall of 1986.

A four-day symposium commemorated the event, marked especially by Prince Charles of Great Britain speaking on the theme of science and society. In paying respects to America's oldest university, the prince chose to criticize the nation's high interest in technology, and in so doing he cited the wisdom of the ancient Greeks. One key sentence from his speech appeared on the front pages of newspapers the next day: "We may have forgotten that when all is said and done a good man, as the Greeks would say, is a nobler work than a good technologist."

The linking of Greek philosophy, the British aristocracy, and Harvard in a rebuke to American technology provides several exquisite ironies.

First of all, the prince presented an exceedingly narrow interpretation of the Greek attitude toward technology. It is true that in the first half of the fourth century B.C. many Athenian intellectuals held technological activity in low regard. Plato made this clear in several of his dialogues, and his contemporary, Xenophon, stated the case bluntly: "What are called the mechanical arts," he wrote, "carry a social stigma and are rightly dishonored in our cities." But this view was held by a limited group at a particular moment in classical antiquity. Two hundred years before Plato, the Athenians authorized Solon, their chief magistrate, to initiate economic and constitutional reforms, and Plutarch tells us that as part of his scheme to design a stable and prosperous society, Solon "invested the crafts with honor." "At that time," says Plutarch, "work was not a disgrace, nor did the possession of a trade imply social inferiority." A century and a half later, technology was still treated with high regard, as Sophocles confirmed in the famous chorus from *Antigone:* "Wonders are many, and none is more wonderful than man; the power that crosses the white sea, driven by the stormy south-wind, making a path under surges that threaten to engulf him; . . . turning the soil with the offspring of horses, as the ploughs go to and fro from year to year . . ."

By Plato's time, however, the situation had changed radically.

High culture had come to Athens with a vengeance. The ideal of the new Athenian citizen was to care for his body in the gymnasium, reason his way to truth in the academy, gossip in the agora, and debate in the senate. The design and manufacture of material things was not deemed worthy of a free man's time. We may well wonder if this patrician attitude did not contribute to the ensuing decline and fall of the Athenian state.

Platonic scorn for technology has from time to time returned to haunt engineers and nowhere more virulently than among the upper classes of Great Britain. There the landed gentry considered engineering to be an occupation suitable for the lower classes, and this prejudice was reflected and nurtured in that nation's great universities. Although Britain was the cradle of the industrial revolution, neglect of engineering education put the nation at a disadvantage in international commerce by the mid-nineteenth century. In France, Germany and other Continental nations, technical education was socially esteemed and politically supported. It can be argued that the decline of Britain as a world power had its roots in that long-ago expression of snobbery.

Upper-class prejudices against technology were exported from England to the United States and found a home at Harvard and other eminent universities. "Applied science" was grudgingly introduced at Harvard in 1847 with the founding of the Lawrence School—endowed by a manufacturer of woolen goods—but that school graduated only forty-nine men prior to the Civil War, and this according to an engineering educator, "in the face of an unconcealed disdain on the part of the regular faculty."[9] Such disdain doubtless helped convince the founders of MIT of the need for their new institution. If the United States had been guided by Harvard's concept of higher education, we might well have followed Athens and Britain into the backwaters of history. (I discuss these issues at greater length in Chapter 10.)

In all the articles about Harvard that appeared during the week of its celebration, in all the praise for its various departments and for its outstanding graduate schools of business, government, law, and medicine, I was surprised to see no mention of the university's lack of attention to engineering. (In the year preceding the anniversary celebration, Harvard awarded only thirty-eight undergraduate degrees in engineering science plus six masters and eight doctorates.) If MIT were not its close neighbor—making the development of its own engineering school at this late date somewhat superfluous—Harvard would be seen to be a seriously flawed institution.

The appearance of the Prince of Wales at Harvard's celebration, and his harking back to Greece in warning against the evils of technology, made for a pleasant tableau and provided a laudable endorsement of scholarship and virtue. But it also evoked memories of aloof philosophers, complacent aristocrats, and haughty academics, people whose view of the world has been needlessly fastidious and dangerously shortsighted.

## THE SOUL OF A NEW UNION

If Homer wrote about warriors, Shakespeare about kings, Hardy about farmers, and Balzac about merchants, why don't the creative artists of our time write about engineers? Until this happens, there will be no consummation of the spiritual union between technology and the general culture. Until this happens, our technological society will not be philosophically at peace with itself.

In *Literature and Science,* a slim book published in 1963, Aldous Huxley tried to discern ways in which literary artists might come to grips with the accelerating scientific and technological revolution. Contemporary writers, he observed, have shown little enthusiasm for science and less for engineering. Although they have been concerned with "the social and psychological consequences of advancing technology," they have been very little interested in technology itself. The making of machinery has not aroused—in either writers or readers—the "passionate interest" that lies at the heart of creative literature.[10]

Looking to the future, however, Huxley was more intrigued than discouraged. He viewed the difficulties inherent in wedding science to literature as a challenge to intellectual combat. "The conceptual and linguistic weapons," he said, "with which this particular combat must be waged have not yet been invented. . . . But sooner or later the necessary means will be discovered . . ."[11]

As the years have passed, there has been little indication that Huxley's optimism was well founded. Science has inspired numerous works of poetic insight; but technology, increasingly complex and impersonal, appears to be drifting ever further away from the domain of serious literature. There is science fiction, of course, but it is often written as superficial entertainment and usually set in the future, as if to concede that the world of complex machines cannot be congruent with the world of contemporary sentient beings. The Society for Literature and Science, founded in 1985, "fosters the multidisciplinary study of the relations among literature and language, the arts, science, medicine, and technology." However, its journal, *Configurations,* consisting mainly of academic exegesis, sheds light on the problem rather than showing the way to a solution. There is little that commentary can do to fan the flames of creative genius.

There have been a few successful efforts. John McPhee has explored aeronautics and nuclear technology, and in *The Control of Nature* (1989) he considered the "heroic chutzpah" that drives people to challenge mudslides in Los Angeles, floods along the Mississippi, and volcanoes in Hawaii. David McCullough has written two splendid epics: *The Great Bridge* (1972), about the building of the Brooklyn Bridge, and *The Path Between the Seas* (1977), the tale of the Panama Canal. These are to my taste among the very best nonfiction books of our time; but they deal with past grandeur, implying lack of appeal in the engineering of our own age.

The History of Technology has developed into a respected discipline and generated many fine works, but most of them attract limited audiences. Henry Petroski, a civil engineering professor at Duke University, has written several very good books with a popular touch—most notably *The Pencil* (1989) and *The Evolution of Useful Things* (1992)—but they represent an offering from the technical community rather than an embrace by the artistic. The same reservation applies to Steven Levy, a columnist for *MacWorld* magazine and author of *Hackers* (1984) and *Insanely Great: The Story of the Macintosh* (1994), two delightful and informative works, but told by a quasi-technical insider rather than an artist observer.

Turning to creative literary figures, I think first of Alexandr Solzhenitsyn. In his novel, *The First Circle* (1968), he showed that engineers at work can be the subject, however tangential, of powerful fiction. Robert Pirsig's *Zen and the Art of Motorcycle Maintenance* (1974) blazed across the literary sky, albeit eccentrically like a comet that might not return for a hundred years. There are other instances, for example Norman Mailer and Tom Wolfe bemused by America's early exploits in space. But aside from a few such tantalizing glimpses, technology is largely absent from contemporary literature, except of course—as with Thomas Pynchon, Kurt Vonnegut, and many others—where it serves as a foil for satire or a backdrop for scenes of alienation.

The one marvelous exception to this unhappy state of affairs came along in the summer of 1981. I refer to Tracy Kidder's book, *The Soul of a New Machine,* a unique work that goes far toward fulfilling Aldous Huxley's optimistic prophecy.

On its face the book appears to be an account of how, from mid 1978 to early 1980, a group of engineers at Data General Corporation developed a new 32-bit supermini computer. But Kidder endows the tale with such pace, suspense, and excitement that he elevates it to a high level of narrative art.

Beyond the narrative, or rather woven into it, is the computer itself, described physically, mechanically, and conceptually. The descriptive passages will not "explain" computers to the average reader, but they give a feeling, a flavor, that adds to one's "understanding" as broadly or even poetically defined.

Kidder proceeds by taking the reader "down into" the machine, and indeed the book consists of repeated descents—not only into an environment of wires and silicon chips, but also into dark corporate basements where secret work proceeds feverishly behind locked doors, and into home cellar workshops where engineers pursue their compulsive tinkering. One of the senior engineers introduces Kidder to the game "Adventure" in which the computer appears to create an underground world called Colossal Cave through which the player must travel by typing out directions on a terminal keyboard. This world consists of mazes, twisting passages, dark chambers, and rusty doors; it is populated by dragons, snakes, and trolls, all creations, of course, of the computer engineers who invented the game. Reading this book is, in part, a voyage through such a subterranean world. Kidder is our Dante—not, to be sure, a mature genius artistically reconstructing Western civilization at the end of an era, but a venturesome young explorer standing on the threshold of a new age looking for the outlines of uncharted regions of human experience.

His companions in this journey are a cadre of about two dozen engineers, who, working day and night under incredible pressure for almost two years, produced the new machine code-named Eagle. These characters, introduced in succession as their role in the unfolding drama becomes significant, are surely drawn larger than life; but this is totally appropriate in a work of imagination cast in the form of a journalistic report.

The leading lights of the Eagle team, chosen for their brilliance, energy and ambition, are portrayed as eccentric knights errant, clad in blue jeans and open collars, seeking with awesome intensity the grail of technological accomplishment. Practically all of them, we learn, were obsessed from their earliest years with the need to see *how things work,* taking gadgets apart and putting them back together. In technical creativity they have found a fulfillment that occasionally verges on ecstasy—"The golden moment . . ." "When it worked I'd get a little high . . ." "Almost a chemical change . . ." "It was the most incredible, soaring experience of my life . . ." By plunging into the world of numbers, theories, and things, they appear to find a path to their own emotions. By looking outward they reach inner depths. In *doing* they encounter *being.* The contrast with the narcissism of most contemporary fiction is striking. Wives and children drift occasionally across the background, mellow and serene, as if intense interest in one's work were the key to domestic felicity. Again the contrast with contemporary literary clichés is remarkable.

The Eagle team is divided into two working groups, "the Hardy Boys," who put together the machine's actual circuitry, and "the Microkids," those who develop the microcode that fuses the physical machine with the software programs that eventually tell it what to do. These men—none of the engineers is a woman, in spite of equal opportunity recruiting efforts—these men, then, are fanatics but not purists. They cannot afford to be; it is crucial that they not only produce a superior machine, but also work quickly enough and cheaply enough so that it will "get out the door" to market. The most elegant technical solution is worth nothing if the end product is not *used.* This need to stop striving for perfection—to say at some point "OK, it's right. Ship it."—is a bittersweet aspect of the engineering experience that has implications for other elements of our public and private lives.

Being totally absorbed in their work, the Eagle engineers are vulnerable to exploitation, and Kidder describes in detail the often devious means by which the members of the group are recruited and persuaded to "sign up," by which is meant not merely enlisting, but throwing one's entire being into the enterprise. The men put up with cramped quarters, inadequate supplies, unpaid-for overtime, moody often uncommunicative bosses, and in the distance somewhere, corporate overlords known to be ruthless and aggressive—yet morale remains surprisingly high. One young engineer, burned out and feeling the pressure in his stomach, leaves suddenly, announcing that he is going to a commune in Vermont. But the others persist, grumbling and weary, yet perversely playful and tenacious, arguing constantly, yet working "in sync." The project becomes a crusade.

At a time when American productivity is an issue of concern, when the nation's innovative powers are said to be waning, and nobody seems to be able to motivate himself or anybody else, the experience of the people who created Eagle merits attention. Not that life can be lived in a state of perpetual commotion. But in microcosm the Eagle team exhibits the intensity and high spirits that social commentators keep saying Americans have lost.

The leader of the group is an engineer named Tom West, who is introduced in the Prologue at the helm of a small white sloop sailing in rough seas. Quiet, aloof, and intrepid, West is described by one of his sailing companions as "a good man in a storm." (I could not help thinking of John Hersey's novel, *Under the Eye of the Storm,* in which the computer scientist, flawed by an "electrical intellectuality," disintegrates during a crisis at sea, while the hero, "a humanist, a vitalist," performs valiantly.) In his youth, West had been required to leave Amherst College for a year as "an underachiever," and almost became a guitar-playing dropout. But he responded to the chaos around him—it was the early 1960s—by deciding to become an engineer. His friends were astonished: "The very word, *engineer,* dulled the spirit." Yet West felt that "in a world full of confusion there is satisfaction to be found in learning how things get put together, how they work." By 1978 he was at Data General in charge of the Eagle team, an austere demanding Ahab of a commander who led his young crew in chase of a contemporary white whale. The hunt was successful, except that after the triumph and the glory came the tragic recognition that for each individual the quest must start afresh, and that life may never again be as exciting.

For all his dramatic flair, Kidder never engages in cheap promotion of the computer mystique. Near the end of the book, with the successful conclusion of the project in view, he joins the Eagle group on a day's excursion from their Westborough, Massachusetts headquarters to a computer trade show in New York. After looking through the exhibits the young men scatter throughout the city to enjoy an afternoon's relaxation. Kidder sits in a café with one of the engineers and looks out at the crowds and the traffic. He realizes that computers, for all their magical qualities, are not about to change the essence of the human condition. And as the computer engineers return to their bus, bubbling over with the effects of their holiday and a few beers, the reader cannot help concluding that the imminence of a sinister technocracy is one of the silliest myths of our time.

Since *The Soul of a New Machine* was published in 1981, I have been waiting for another work in the same genre, another work in which engineering and technology are imaginatively depicted as part of the human scene. Tracy Kidder has gone on to other themes: housebuilding, teaching, and care of the aged, by which he has demonstrated a determination to consider some of the most vital issues of our time. It is meaningful that he started with the world of engineering. It is high time that others follow where he has led. While we wait, however, it is good to know that the way has been shown. I believe that Aldous Huxley—who looked forward to the coming of a worthy literature of science and technology—would be pleased.

# II

# ENGINEERS

CHAPTER 7

# WHAT IS AN ENGINEER?

We can examine engineering indefinitely without coming close to exhausting its possibilities, its meanings, and the ways people feel about it. Such study and debate is an appropriate exercise for anyone, not just engineers. But discussion of *engineering,* and its place in the world, can only tell us a limited amount about *engineers.* It is time that we moved from the macrocosmic overview of technology to the microcosmic consideration of engineers and their profession.

## BEETHOVEN'S RESPONSE

What is an engineer? I have tried to answer this question from time to time in various writings and speeches, as have many people both within and without the profession. Yet a satisfactory definition remains elusive. Perhaps the best response is the one attributed to Beethoven when he was asked a question to which he was unable to formulate an adequate answer. According to the story, after having performed one of his piano sonatas, the great artist was asked by a listener to describe what the piece "meant." After a moment's thought, Beethoven turned back to his piano and played the sonata again.

Taking a lead from Beethoven: Engineers are what they are. And, to a great extent, engineers are what they do. The ultimate definition of an engineer is to enumerate all the products and processes created by engineers and to describe the activities by which these products and processes are achieved.

In the room where I am writing at the moment there are lamps, a mirror, a television set, a telephone, carpeting, painted walls, and the computer that serves as my word processor. I am in a house. Outside I see automobiles, and overhead there goes an airplane. Happily, I am overlooking water and can see motor boats, sailboats, and a fishing trawler. Every manufactured object has been designed by engineers, and its manufacture planned and overseen by engineers.

In the United States in the mid-1990s there are about 2.5 million people described as engineers by the National Science Foundation, mainly because they have received degrees from accredited engineering schools. About 1.5 million of these actually do engineering work as defined by the Bureau of Labor Statistics. The others do selling, business management, and a myriad other jobs, or are out of the labor market altogether.

Engineering specialties, based upon current undergraduate enrollment, break down as follows:[1]

| | *Percent* |
|---|---|
| Electrical and Computer | 31 |
| Mechanical and Aerospace | 26 |
| Civil and Environmental | 16 |
| Chemical and Petroleum | 10 |
| Industrial and Manufacturing | 5 |
| Other (Marine, Mining, Materials, Nuclear, etc.) | 12 |
| | 100 |

Classified by type of activity, the profile for engineers shortly after graduation looks like this:[2]

| | *Percent* |
|---|---|
| Research | 5 |
| Development | 30 |
| Management of R&D | 3 |
| Production, Operations, Quality Control | 27 |

| | |
|---|---|
| Other Management | 10 |
| Statistical Work, Computer Applications | 8 |
| Sales, Professional Services | 5 |
| Teaching | 3 |
| Other | 9 |
| | 100 |

For a more comprehensive explanation of what engineers do, I know of no better guide than James L. Adams's book, *Flying Buttresses, Entropy, and O-Rings,* published in 1991. In addition to the above classifications by "field of study" and "type of activity," Professor Adams notes other significant subdivisions:

By industry: Electronics, automotive, aerospace, etc.
By "discipline" (which often crosses several "fields of study"): Heat transfer, fluid mechanics, structural analysis, control system theory, circuit design, etc.
By product: Integrated circuit, spacecraft, laser, lawnmower, etc.
By industrial process: Plating, ultrasound, refinery, diffusion, etc.
By responsibilities: City, military, air pollution control, etc.

Most engineering work, whatever the specialty, is done in a process that has a series of sequential stages. As Adams puts it: "The engineering process begins with a desire. This is reduced to a problem."[3] Definition of the problem is usually followed by "preliminary design" of a product, and then by "detailed design." Next comes "development," including testing, to bring a prototype to a functional and economic level that is deemed satisfactory. (The product must "work" well and also be produced at an acceptable cost.) The last step in the process is "production," although contemporary thinking stresses planning for production in the early stages of design. Manufacturability, which once was something of an afterthought, is today very much in the forefront of the engineering process.

Along with a finished product there are considerations of marketing, delivery, maintenance, and service. And in this age of environmental sensitivity (and environmental laws), disposal of products and byproducts of engineering work are critical concerns. Finally—or perhaps foremost—underlying the entire engineering initiative, is R&D, research and development, those investigations that are carried out specifically to produce new knowledge and new products.

The ability to do engineering design derives initially from inner vision, from a sense of how things "fit" and how things "work." Eugene Fergusen, in an estimable book called *Engineering and the Mind's Eye* (1992), has written of the need for "sound judgment and an intuitive sense of fitness and adequacy."[4] However, good sense and conceptual facility constitute only a small part of professional engineering at the end of the twentieth century. Instinct and craftsmanship have been enriched with the findings of science. Contemporary engineers are students of "the engineering sciences": materials, structures, fluids, electricity, light, heat, energy, chemicals, systems—all the phenomena that constitute the physical universe. Underlying these engineering sciences are three fundamental disciplines: mathematics, physics, and chemistry—and, increasingly, a fourth: biology.

By contemplating the many subjects that engineers must study, by considering the stages of the engineering process, and by itemizing the various specialties through which engineers organize themselves, we begin to develop some understanding of the profession. This enables us to circumvent the fruitless search for a comprehensive definition of the word *engineer.*

Nevertheless, there are those who will not give up the quest, who seek to abstract overriding verities from the complexity of real life. A most notable effort along these lines is the book, *What Engineers Know and How They Know It* (1990), by Walter G. Vincenti, an eminent aeronautical engineer who made many contributions to the development of aviation. Doctor Vincenti attempts to define "the character of engineering knowledge as an epistemological species." He modestly concedes that his "model for knowledge growth" is "relatively conjectural and subject to controversy." Nevertheless, for all its puzzling complexity, his presentation reveals additional insights, and a heightened awareness of what engineering is all about. (On the other hand, I don't want to lose *my* readers at this point, and some of them may wish to skip the next five paragraphs!)

Vincenti begins by calling attention to five problems that had to be dealt with in the early days of aeronautics. These were: (1) analyzing "flow," the phenomenon that underlies the essential understanding of "lift"; (2) designing airfoil shapes (finding the specific wing shapes that are most effective); (3) designing propellers; (4) designing and producing flush riveted joints for aircraft (a "detail" that was crucial to success); and (5) designing controls to obtain flying qualities satisfactory to pilots (the "feel" of the product to the user).

Through historical reviews of how these problems were addressed and solved, Vincenti demonstrates many of the ways in which engineering progress occurs—a rich mix of theory and experimentation, intuition and craftsmanship, mathematics, drawing, modeling, and luck. (Much of the experimental work done in the early days of aviation would be replaced by computer modeling today.) He demonstrates how uncertainties of knowledge plus exigencies of the marketplace required activities to be performed in parallel that should logically have been done in sequence.

Vincenti then identifies seven "knowledge-generating activities" in which engineers engaged while addressing the problems he has delineated. These are, in his terminology: Transfer from science; Invention; Theoretical engineering research; Experimental engineering research; Design practice; Production; and Direct trial. He suggests that these activities yield six categories of "engineering design knowledge": Fundamental design concepts; Criteria and specifications; Theoretical tools; Quantitative data; Practical considerations; and Design instrumentalities. Vincenti admits that such categorization is "not usual in the profession," thus offering solace to any reader who may not follow his hypothesis in all its detail.

In a final chapter, entitled "A Variation-Selection Model for the Growth of Engineering Knowledge," Vincenti explores the concept of growth in engineering knowledge through blind variation plus selective retention, sort of a Darwinian process. According to this theory, many possibilities pop into an engineer's mind through blind chance, and then are screened. Only the fittest survive. This is clearly one way in which engineering creativity occurs, although few engineers would agree that it is the only way, or even the main way.

In addition to theory, Vincenti demonstrates that engineering is flesh and blood, real people overcoming uncertainties and frustrations. He also reminds us of the importance of organizations in technological progress. In the development of aviation, major roles were played by the National Advisory Committee for Aeronautics (set up by Congress in 1915), university engineering departments, aircraft manufacturers, the military services, airlines, and professional engineering societies. There were also what Vincenti calls informal "communities of practitioners"—for example, a "fastener community" involved in the development of flush riveting—which were crucial to technical progress. The importance of the group in engineering work was already apparent a generation and more ago.

Even after we describe what engineers do—with or without attempts at comprehensive exegesis—there still remains untouched an important aspect of what, or who, engineers are. There is an entire realm of personality, lifestyle, and philosophy that is not defined by job description. Being an engineer entails looking at the world in a distinctive manner, and experiencing the world in singular ways. In my book, *The Civilized Engineer* (1987), I attempted to set forth what I took to be the main elements of "the engineering view," a general outlook, a way of approaching the world, that is shared by members of the engineering profession. I summed up my thoughts this way:

> These, then, are what I take to be the main elements of the engineering view: a commitment to science and to the values that science demands—independence and originality, dissent and freedom and tolerance; a comfortable familiarity with the forces that prevail in the physical universe; a belief in hard work, not for its own sake, but in the quest for

> knowledge and understanding and in the pursuit of excellence; a willingness to forgo perfection, recognizing that we have to get real and useful products "out the door"; a willingness to accept responsibility and risk failure; a resolve to be dependable; a commitment to social order, along with a strong affinity for democracy; a seriousness that we hope will not become glumness; a passion for creativity, a compulsion to tinker, and a zest for change.[5]

In an earlier book, *The Existential Pleasures of Engineering* (1976), I argued that engineering is an occupation that responds to humanity's deepest impulses, and is rich in spiritual and sensual rewards.

Clearly this discussion of the word *engineer* has begun to drift into a euphoric mode. Indeed, each time I start to talk about engineers and the work they do, I have to guard against getting carried away. It all seems so *interesting,* so important, so gratifying—so existentially fulfilling—that I find myself speaking in hyperbole, a very un-engineering way of behaving.

## RANK AND FILE

I fear that like many people who write about "the engineer," I tend to think in terms of a composite individual who graduates from engineering school, enters the ranks of the profession, and embarks on a career filled with challenge and opportunity. This person may be of indeterminate sex or color, but will predictably be proficient, inquisitive, and ambitious. Naturally, not all engineers can be equally happy and "successful," but I'm inclined to assume that this is merely a question of degree.

The archetypical engineer of my imagination creates, discovers, and produces, all as depicted in the handsome magazines published by the major engineering societies. These periodicals feature impressive technical achievements plus inspiring individual success stories, presented in stylish layouts with multicolored illustrations. They confirm for me a glowing image of the profession.

Reality, however, presents us with complexities that cannot be accommodated within such a pleasant paradigm. The practice of engineering is not simply a matter of starry-eyed progress for each individual—onward and upward, as it were. There is a grimmer aspect to the profession, one that is seldom addressed in the literature of technical journals. This dark side of engineering reveals itself from time to time in the letters written to the professional societies by disenchanted members.

One such anguished and angry letter appeared in *The Institute,* a publication of the Institute of Electrical and Electronics Engineers. "The IEEE doesn't represent the needs of the majority of working engineers," declared the correspondent, who had spent thirty-five years as a design and development engineer in radar systems and high-energy accelerators. The organization "adopts an elitist stance of what's good for academia and corporate management," and as a result "only a small fraction of my peers have been or are members."[6]

It is true that of the approximately 2.5 million American graduate engineers—about 1.5 million of whom, as I have said, are engaged in engineering work—only 750,000 are members of any of the major engineering societies. It is also true that no more than a third of engineering graduates engage in studies beyond the four-year bachelor's degree; and less than 20 percent bother to obtain professional licenses. Whatever meaning one cares to give to these figures, they do not tell a pretty story.

A study of engineers in the Rochester, N.Y., area found that almost one third do not share the commitment to work values that one ordinarily associates with engineering. They do not expect to be involved in exciting new technical developments nor do they aspire to significant personal success. They seek merely "a modest level of technical challenge combined with the opportunity for periodic promotion and their share of organizational recognition." They can best be described, according to the authors of the report, as a kind of engineering "rank and file."[7] When I spoke about this study to a group of engineers who are active in, and concerned about, professional affairs, several thought the findings very discouraging. Indeed, there is much about engineering that can discourage advocates of the profession once they take off their rose-colored glasses.

Yet the history of engineering reveals many important contributions—individual as well as collective—made by underlings and independents, even by malcontents, and especially by mavericks. It will not do for leaders of the profession merely to deplore the lack of commitment that appears to be so widespread. It is important that they empathize with the concerns of the alienated and make special efforts to bring them back into the professional family. It is the way of the world—and not just for engineers—that as professional leadership accrues to energetic "organization" people, movement toward democracy tends to languish.

Clearly there is no archetypical "engineer" about whom we can make sweeping generalizations. From the "rank and file" to the most productive creators, from the angry and alienated to the inspired and committed, the profession contains a wide variety of human types—as well it should, engineering being an elemental expression of the human spirit.

## HARD TIMES IN THE FOREST OF ARDEN

It is sobering to recognize that not all engineers are the heroic figures of my sometimes fervid imagination. It is doubly sobering to find that in the mid-1990s engineers have had serious problems getting and keeping jobs.

Engineering unemployment in the United States, which for several decades averaged about 2 percent, in 1992 suddenly jumped to about 4 percent, and stayed there through 1995, showing little indication of improvement in the near future. Although this is still far below the general unemployment rate in the United States, the increase came as quite a shock and brought much anxiety. (Even in the oft-recalled "crisis" year of 1971, the rate was only 2.84 percent, and improvement from this figure came swiftly.)

Also, the magnitude of the problem goes far beyond the 4 percent figure. R. A. Ellis, Director of Research for the American Association of Engineering Societies, estimates that at least 10 percent of the engineering workforce has been at some time unemployed, most engineers finding new positions when discharged from old, but still undergoing considerable trauma.[8] Further, the impact on engineering students has been dispiriting. Some prominent schools report more than 40 percent of their graduates without employment at graduation, and half of these still unsettled six months later, much higher figures than ever in recent memory.

The main reasons for this decline are readily apparent: sluggish economic conditions; fierce international competition which prompts "downsizing," "outsourcing," and cutbacks in industrial R&D; congressional reluctance to fund "unnecessary" projects, such as space exploration; and, perhaps foremost, cutbacks in the military budget following the end of the Cold War. (Up until 1990, 20 percent of the nation's engineers worked in the defense industry, with perhaps another 10 percent involved in products used partly by the military.) Knowing the reasons does not lessen the pain. And painful it has been.

Yet "Sweet are the uses of adversity." The words are from Shakespeare's *As You Like It,* spoken by the good Duke Senior, whose kingdom has been usurped by an evil brother. Trying to cheer up his fellow exiles, the Duke extols the benefits of their rigorous life in the Forest of Arden. I consider the phrase apt because of other news about engineering employment—good news that follows hard on the heels of the discouraging figures just recited.

Admittedly, no abstract social improvement can be balanced against the immediacy of hurt to individuals. But having said this, one can see potential benefits in the current crisis, some of them quite exciting. The good news is summed up in a headline in the professional periodical, *ASME News:* "Engineers Are Finding New Directions for Skills."[9] The article relates how Career Transition Workshops have helped engineers recognize the many occupations for which their talents and experience have qualified them. Some engineers have always been adept at switching specialties, and some have traditionally found their way into sales, management, and just about every other career path imagineable. But adversity has helped sharpen their awareness of the breadth of their competence.

At the same time, completely new sources of employment have begun to appear. These stem partly from technological innovation (the classic hoped-for solution to technological unemployment), but also from a heightened appreciation of what engineers can do. Banks, for example, are recruiting engineers because of their knowledge of information systems. Investment firms and financial houses of all sorts are seeking out engineers to do research, and because of their general quantitative and problem-solving skills. Computers have transformed the world of securities trading, with engineers contributing to increasingly sophisticated modelling techniques.

Andersen Consulting, the information systems consulting arm of the Arthur Andersen accounting firm, has started employing engineers by the hundreds. Management consultants are also looking to engineers, as evidenced by the hiring activities of McKinsey & Company, Inc. Environmentalism, an ever-growing complex of concerns and activities, has brought corporate and government agency recruiters to the campus. According to the *Engineering Workforce Bulletin,* this helps to explain why chemical engineers have been getting jobs in spite of weak demand from the chemical industry. Civil engineers are also equipped to work in the environmental field, and in addition have benefitted from an accelerating concern for the national infrastructure.

While large corporations have curtailed their hiring, small companies—leaner, quicker, more responsive to market changes—have shown gratifying resilience. According to Robert K. Weatherall, MIT's Director of the Office of Career Services, placement directors around the nation have noted that these firms are looking for engineers who are "bright, all-round, practical individuals." "Specialists" are out of favor. Indeed, the more specialized branches of engineering—aeronautical, materials, nuclear, ocean, etc.—have suffered the most during the current downturn.[10]

An optimist listening to reports from the employment frontiers will conclude that engineers are exceptional people, qualified to do practically anything. In this increasingly technological world, the engineering profession will endure and, in the long run, flourish. Engineering may yet become the education "of choice" in the United States as it is in many parts of the globe. But this can happen only if all of us—engineering educators especially—are alert to the changing needs of the job market and of society as a whole. The nation needs engineers who are *versatile* in the best and most comprehensive meaning of the word.

Perhaps for American engineers, as for the exiles in Shakespeare's Forest of Arden, the uses of adversity may indeed turn out to be sweet.

## SERENDIPITY

Beyond anxiety about the job market of today, there lie uncertainties that have troubled engineering students in every generation. Inevitably they wonder: Once I become a professional engineer, how do I make the most of it? Whenever I visit the campus of an engineering school, some earnest undergraduate is sure to ask me questions about how best to embark on a career.

It is difficult not to answer in platitudes, which are, after all, the residue of truth.

Firstly, know thyself. Young engineers do not all have the same aspirations. To many, the opportunity to do exciting technical work is a top priority. To others, job security has greater appeal. There are idealists who want to help save the world, while some dream mainly of large incomes. A few have entrepreneurial ambitions. The list of hopes and dreams is endless: teaching or government service; freewheeling independence or position in a reputable corporation; pressure or serenity; a job near home or travel to exotic places; and so forth. A goal, however nebulous, provides a valuable inner compass.

I usually mention my predilection for taking chances when one is young. There is time enough for caution when family responsibilities begin to accumulate or the fires of energy burn a little less brightly. I counsel patience: Do not expect the first days of employment to be blissful. On the other hand, I often quote Charles Steinmetz, the brilliant electrical engineer of an earlier age, who said that young engineers who find themselves in dead-end jobs should seek to escape, just as a drowning person will seek the surface of the water "to get the blessed air."



Inevitably the student will ask about my personal experience, and this gives me the chance to switch the conversation from planning to serendipity. For it is my strong belief that in the pursuit of a career, luck, timing, and pure happenstance are likely to be critical. Engineers should expect this and be prepared to make the most of it. I do not counsel preparing to give up on one's dreams, rather being flexible, imaginative, and willing to consider alternate routes to professional satisfaction.

When I first studied structural design, I thought that I had found my life's calling; but this was not to be. I completed my engineering studies in the Navy, and at the end of World War II found myself helping rebuild military facilities on various Pacific islands. Presto, I became a construction engineer instead of a designer. Upon returning to civilian life, unappeased wanderlust led me to a surveying job in South America, after which I resolved to forever spend my working days outdoors, "in the field." However, upon returning home, the best available opportunity was as an office engineer doing estimating and project management. Weighing the alternatives, I traded in my transit for a pencil, a calculating machine, and a telephone. This worked out wonderfully well, except for one little problem: the venerable company I joined—builder of many of New York's most outstanding landmarks—unbeknownst to me, had for some time been experiencing financial difficulties. As I rose in the ranks, the firm descended toward insolvency.

Eventually I encountered two young men who had just started a contracting business in Westchester County, a half hour "reverse commute" from my home in Manhattan. At that very moment (timing is all) they needed what I had willy-nilly become: an estimator/project manager with field experience. I threw in my lot with this new enterprise—and found, at last, my vocational utopia.

I would guess that my career path is no more erratic than that of most engineers, particularly in recent times. The technological world is more mercurial than it has ever been. Companies come and go, and products flourish only to become outmoded. Happily, engineers, unlike dinosaurs and other extinct species, seem infinitely adaptable.

As the archetype of the serendipitous engineer, I submit the example of a young man I know who majored in petroleum engineering, thinking that one day he would uncover new energy sources under deserts or ocean floors. Upon graduation he had trouble finding work in the oil industry. But opportunity came from a new and totally unanticipated quarter. Today he is very successful—and professionally fulfilled—analyzing and remediating soil under abandoned gasoline stations.

This is the sort of future I try to project before young engineers. Make plans, to be sure, and things may work out exactly the way you anticipate. But if they don't—and they probably won't—be thankful for the marvelous variety that is inherent in your profession.

## DANGERFIELD'S COMPLAINT

Indeed, there are many reasons to be optimistic about the profession, and to expect the future role of the engineer to become increasingly important. But changes in professional and social status tend to come slowly. For the moment, anybody who has spent time with engineers knows that they are disgruntled about their place in society as compared to other professionals.

Wherever engineers gather one is likely to hear complaints about lawyers, politicians, bankers, and MBAs, all of whom seem to have power, wealth, and renown beyond their true deserts. It isn't an attractive stance to take, being a malcontent, sort of like the comedian, Rodney Dangerfield, who protests about getting "no respect." Yet, when one considers how engineers are treated in the media, their resentment can be better appreciated. For example:

When Paul Lego, the chairman of Westinghouse, retired under pressure in 1993, *USA Today* characterized him as "a cerebral engineer," ill at ease with those aspects of the firm's business that were not in the traditional nuts-and-bolts mold. When Robert C. Stempel, the chairman of General Motors, resigned the previous year, the *New York Times* pointed out that he was "the first engineer in decades" who had held the top post, and that he was unable to cope with such nontechnical challenges as a union strike and an impatient board of directors. *Fortune* quoted a source who knows GM intimately as saying, "Stempel is a dynamite engineer, but he should never have made it to the top of the ladder."[11]

When a safety engineer was called upon to testify in court, a newspaper reporter characterized him as having "the stolid manner of an engineer."[12] When a university president, who happened to be an engineer, was forced to resign because of claims of sexual harassment, the *Philadelphia Inquirer* made note of his efforts to emerge from the emotionally guarded life of an engineer.[13] And so it goes. Perhaps it takes a touch of paranoia to find a pattern in references like these. But engineering paranoia—and I do not deny that it exists and I suffer from it—comes only after one has seen the profession slighted time and again.

More serious than the occasional media caricature of the engineering personality is the constant recognition paid to scientists, often for work that is done by engineers. A particularly irksome example had to do with the *Challenger* disaster in 1986.

As the ill-fated shuttle was readied for takeoff, stories in the *New York Times* referred to NASA "officials" and "spokesmen," and discussed planned "science experiments," with nary a mention of engineers or engineering. That same week, *Voyager 2* was sending back dazzling pictures from the planet Uranus, and *Times* reporters waxed rhapsodic about the event. In front-page articles they spoke of "science," "project scientists," and "scientific" achievement, nowhere referring to the engineering miracle of designing that craft and sending it out into space.

The day after the *Challenger* disaster, the *Times* was suddenly filled with discussion of failed "technology." In a remarkable shift in terminology, it was now "space agency engineers" who would try to diagnose the cause of the catastrophe, "engineers" at Mission Control who were saddened and feeling a sense of failure.[14] In the hearings that followed, the theme most frequently sounded was that of engineering responsibility. The average reader is not likely to notice such nuances, and perhaps each incident is of little import. Nevertheless, the cumulative effect can be consequential.

The good image of science is related to the purity of the search for truth, whereas the ambiguous image of engineering is related to the making of things. Thus it seems particularly unfair that when scientists become engaged in making things—that is, when scientists do engineering—the credit so often goes to science. A case in point is the continuing saga of superconductivity. Although the relevant theoretical concepts may fall mostly in the domain of physics, the fabrication and testing of materials can truly be called engineering. Yet when the topic is mentioned in the media the talk is mostly of science, scientists, and scientific breakthroughs. On the other hand, it is interesting to find an article on the subject in *The Institute,* published by the Institute of Electrical and Electronics Engineers, that does not once use the word "science." Referring to "reports," "findings," and "research," it ends up quoting a professor of electrical engineering![15]

One must look far and wide for examples of neutrality in this clandestine war of words, but they do exist. An article in *Science* magazine dealt at length with superconductivity and managed not to mention either science *or* engineering. The terms used were "researchers," "theorists," and "experimentalists."[16]

Some may consider all this a bit silly, and admittedly there are more pressing problems before us. But when one considers the competition between professions—for research funds, talented students, public support, and much else—it becomes clear that more is at stake than mere pride. (And who is to say that pride is not important?)

It is some consolation that whenever polls are taken to gauge the public's opinion of various professions, engineers usually earn high marks. Most of the public has a "generally high opinion" of engineers, and admires their honesty and ethical standards.[17] Yet, gratifying as this may be, it does not compensate for the lack of status that engineers perceive reflected in the media.

## THE ANONYMOUS PROFESSION

One thing we can safely say without being accused of paranoia is that practically none of today's engineers become famous. That's not the worst thing in the world, of course. Most poets are not famous either. Has anybody else noticed that engineers and poets stopped being famous at about the same time? Herbert Hoover—the last engineer who achieved world-wide fame—died in 1964, the year after Robert Frost and the year before T. S. Eliot—the last poets whose names were household words (at least in some households). If Hoover, Frost, and Eliot don't meet one's criteria for greatness, it could be said that the last engineers of renown died off in the 1920s—I refer to Eiffel, Steinmetz, and Roebling—a decade before the last poets of international eminence—Kipling and Yeats. This shared loss of fame, this slipping into a mist of anonymity, is a link of sorts between engineering and the arts, although not one in which any of us can take much comfort.

Modern poets, one might argue, have become aloof and precious, retiring from the public scene in what appears to be a willful act of intellectual snobbery. This seems also to be the case with many modern painters and composers. But this is the age of technology. Engineering is where the action is. And yet . . .

Sometimes the anonymity of engineers verges on the incredible. For example: Who invented the integrated circuit? Yes, the microchip that launched the electronics revolution and heralded the coming of a new age. Who invented it? I ask the question rhetorically to make the point that hardly anybody knows the answer—"not one American in ten thousand," according to one historian of technology.[18] It is a truism that scientists and engineers rank low in public recognition. The celebrities of our culture are mostly entertainers, newscasters, and politicians.

We are so accustomed to this state of affairs that we hardly pause to wonder why it exists. The public obscurity surrounding the inventors of the integrated circuit—and other notable technologists—is a phenomenon that practically cries out for scrutiny.

On July 24, 1958, Jack Kilby, an engineer newly hired at Texas Instruments in Dallas, wrote in his lab notebook what came to be known as The Monolithic Idea: "The following circuit elements could be made on a single slice: resistors, capacitor, distributed capacitor, transistor." In retrospect the idea seems exquisitely simple. But since, at the time, silicon was not thought a suitable material for making resistors (carbon was the standard) or capacitors (metal and porcelain was better), the idea of making all the parts of an electric circuit out of one tiny piece of silicon was truly a conceptual breakthrough. Within two months, Kilby's idea was made manifest in a working model.

In January of the following year, Robert Noyce, a physicist at the recently formed Fairchild Semiconductor Corporation in Silicon Valley, described in his notebook the main elements of the integrated circuit. His moment of insight came slightly later than Kilby's but was more complete in that it envisioned the various components connected by imprinted metal strips. Although Kilby was first to think of building all the parts of a circuit in a monolithic chip of silicon, he lagged in solving the problem of interconnection; his first models used tiny gold wires inserted by hand. A ten-year court fight over patent priority was eventually won by Noyce, but there was wide agreement in the scientific and engineering community that the two men deserved joint credit for the achievement. They are generally regarded as "coinventors" of the chip, and each has given generous praise for the accomplishment of the other. Both were awarded the National Medal of Science, both were inducted into the National Inventors' Hall of Fame, and jointly they were the first recipients of the Draper Prize, established in 1991 as the equivalent to a Nobel Prize for Engineering. Both men benefitted financially from their work, although Noyce to a greater extent because he was a principal in his firm whereas Kilby was an employee. (Noyce went on to make a fortune as a founder of Intel Corporation.)

The momentous consequences of what these two men accomplished—nothing less than a second industrial revolution—is demonstrated around us each day. Yet practically nobody knows their names. I ask people all the time—including people I would expect to know—and invariably I am greeted with a blank stare.

This is all the more remarkable when one considers that Thomas Edison, after he perfected the lightbulb, was arguably the best-known person in the world. At the age of seventy-five, long past his years of productive work, he still ranked as the "Most Admired American," according to a *New York Times* survey.[19] Alexander

Graham Bell and Henry Ford also became culture heroes as did many other inventors and industrialists of the past.

What has changed? Why are the greatest technologists of our time so shrouded in anonymity? I have posed this question to the same people who fail to identify Kilby and Noyce, and their answers fall into several categories.

First there is the matter of personality. Edison was a grand eccentric, it is claimed, whereas today's inventors are a bland lot who carry pocket calculators and fade into the woodwork. It is true that Kilby (who, by the way, headed the team that *invented* the pocket calculator) is a quiet, unassuming engineer who is happiest working alone in his laboratory. Noyce, however, before his untimely death in 1990, was an extrovert who skied, flew a plane, and sang madrigals. As a successful entrepreneur and a leading spokesman for the semiconductor industry (who associates called "the Mayor of Silicon Valley"), he shared a special spotlight with such flamboyant individuals as Steven Jobs and David Packard. But even these heroes of the marketplace have not achieved more than a fraction of the fame that was accorded the likes of Edison, Bell, and Ford. No, personality is not the key to the problem. Neither is wealth: Packard has been listed among the ten richest Americans. (Bill Gates, cofounder of the Microsoft Corporation and America's richest citizen, is a college dropout who engineers cannot properly claim as one of their own.)

Nor does the explanation lie, as some suggest, in the group activity that characterizes contemporary engineering. It is true that large and anonymous teams inhabit the laboratories of industry, government, and academe. But all the more exciting, I suggest, are those rare feats of individual inspiration such as those of Kilby and Noyce.

Some of my respondents see the problem as one of public relations, or rather the lack of public relations, in an increasingly professional—hence outwardly modest—scientific and engineering community. More TV exposure would help, I agree, but one must also agree with those media people who say that their lack of interest in technologists is an effect of the prevailing situation rather than its cause.

Technology itself has changed since Edison's time, and some observers find new inventions less enchanting than the old ones. Lightbulbs, suspension bridges, and automobiles, so goes this theory, can be "comprehended" by laymen, qualifying their inventors for human pantheons. Integrated circuits, on the other hand, are "black box" phenomena that nobody really understands, so their creators are not likely to capture the public imagination.

Perhaps we are simply jaded. After hundreds of years of technological advances, with discoveries now accelerating exponentially, we are no longer thrilled by a new breakthrough, no longer inspired by an inventor, however brilliant. I suspect that this comes closer to the mark than any of the other theories. Not that we have lost our capacity for wonder—every sunrise, every newborn baby, shows us this is not so—but rather that we no longer associate technology with salvation. For millenia human societies lived in dread of famine and plague, and to the extent that technology promised relief it evoked wonder and delight. The industrial revolution brought miraculous new freedoms and sensory experiences—travel, communication, cornucopias of consumer goods—and to a naive and newly liberated populace there was every reason to view inventors as heroes. But now that so many of us can take shelter and food for granted, and have a surfeit of TVs, CDs, RVs, PCs, and VCRs, the technologists have lost their messianic appeal.

Not that we have achieved a state of utopian contentment. We still have many vital concerns, and some of these are reflected in names that are widely known, including those of a few scientists and engineers. For a long time our greatest obsession was nuclear war—witness the continuing renown of Robert Oppenheimer, Wernher von Braun, and Edward Teller, "Father of the H-bomb." The celebrity status of Lee Iacocca—engineer/manufacturer/salesman—speaks to our concern about losing our status as the world's premier maker of automobiles. The continuing fame of Jonas Salk indicates how anxious we are to be freed of disease. Anyone who discovers a cure for cancer or AIDS will surely achieve lasting glory.

Society's feelings about engineering may change again, sooner than one might guess, and the names of Kilby and Noyce might yet become the stuff of legend. For the present, however, their lack of renown attests to our mixed feelings about the uses of technology. Their deeds burn nonetheless brightly in the firmament of human achievement.

A number of years ago, at the annual meeting of the American Society for Engineering Education, some whimsical academic arranged a debate on the topic, "Is Engineering an Anti-Social, Anti-Human Profession?" I was on the team debating in defense of the profession, and as part of my argument I suggested that *engineering* absolutely was not antisocial or antihuman—although individual *engineers* sometimes are both. This comment generated a good bit of laughter and vigorous nods of assent, from my opponents as well as from the audience.

Perhaps before going farther it would be wise to examine some of the characteristics of engineers, as people, that exacerbate the problems I have been addressing. There is no sense pretending that the fault always lies elsewhere, in our stars or in our critics. If "know thyself" is still the beginning of wisdom—which the ancients said it was, and I think they were right—then a little self-criticism may be in order.

CHAPTER 8

# FAULTS AND FOIBLES

## LEARNING LESSONS IN WASHINGTON

The appointment of John Sununu as White House chief of staff is already a dimming footnote to history. But when the event occurred in 1988 it generated great interest and excitement in the engineering community.

"Can An Engineer Run Bush's Team?" excitedly asked a lead headline in a professional journal. Yes he can, was the implied response, and proud engineers "look forward to seeing his talents on display in Washington."[1]

A graduate of MIT (bachelor's, master's, and Ph.D.) who worked in industry five years designing heat sinks for transistors and diodes, then became associate professor and subsequently associate dean of the engineering school at Tufts, Mr. Sununu had a most unusual background for an American politician. He followed a career path that is the stuff of hortatory speeches at engineering society meetings but hardly ever found in the real world. His political interests began with work on his local planning board in Salem, New Hampshire, and from there he was elected the town's representative in the state House of Representatives. He served as science adviser to the governor of New Hampshire in 1976 and then ran unsuccessfully for state Senate (twice), and in 1980 for the U.S. Senate (where he ran third out of eleven candidates in the Republican primary). In 1982 he captured the governorship of the state, served three two-year terms, and was poised to win an unprecedented fourth term when he joined forces with the Bush campaign.

Small wonder that the accomplishments of this brainy technologist—extolled by an MIT professor as one of those brilliant students who come along "about once every five years"—should arouse pride and excitement in a profession that has been increasingly dissatisfied with its lack of power and prestige.

In March 1990, *IEEE Spectrum,* official magazine of the Institute of Electrical and Electronics Engineers, the largest engineering society in the world, reported that after a year on the job, Sununu had "silenced many critics who doubted that a Washington outsider could be effective in his position." A feature article entitled, "A Die-Hard Engineer in the White House," asserted the widely accepted view that Sununu had become a role model for the profession.

But in late 1991 Sununu was toppled from power, and the very intellectual brilliance that gained him praise from friends, and grudging respect from foes, was cited as a factor in his ouster. He was smart; but not "street smart." Able to marshall data and understand theorems, he showed little patience for human idiosyncracies, and little appreciation of the need for political give and take. He was not a people person. His failures, just like his successes, seemed to be linked to his background in engineering, where facts rule supreme.

I found it particularly galling that Mr. Sununu's replacement, Samuel K. Skinner—widely praised for his talents as a team player—was a lawyer. Engineers complain that lawyers dominate our political processes. Yet the Sununu misadventure provided a sobering lesson.

This was certainly not the first time we had heard about the engineer's inability to perform in the political arena. Much was said on this theme at the end of Jimmy Carter's presidency. Carter, it will be recalled, spent one college year at Georgia Tech followed by three years at the Naval Academy where he specialized in engineering studies. After graduation he served two years on technical experimental ships, and five years in the submarine service, including work under Admiral Rickover in the design of the earliest nuclear submarine power plants. As president, he was widely criticized for his technocratic approach to politics, not only in his failure to communicate well with the public, but also in his ineffectual relationship with Congress. When his single-term presidency came to an end, commentators placed part of the blame on his engineering background.

Columnist Tom Wicker, for example, noted that Mr. Carter used an "engineer's approach of devising 'comprehensive' programs on this subject or that, but repeatedly failed to mobilize public opinion in their support."[2] A reporter for the *New York Times* observed that Carter "had the sometimes inflexible mind of an engineer."[3] Theodore C. Sorensen, who had served as an aide to President Kennedy, predicted that history would categorize Mr. Carter as the second "engineer president," after Herbert Hoover.[4]

Hoover, like Carter, was intelligent and worked hard; but his single term, ending in the depths of the Great Depression, has been deemed one of the most disastrous in the history of the presidency. Hoover was perceived as an insensitive stuffed shirt, in marked contrast to his affable successor, Franklin Roosevelt.

Harking back to Hoover, who ran for president in 1928 as "The Great Engineer," we reach the origins of the engineer-in-politics story, a rather melancholy American chronicle, we must admit.

Engineer politicians have consistently underestimated the idiosyncratic nature of the human spirit. People, unlike machines, are not generally logical, rational, and predictable. Bismarck is credited with having said that politics is not a science but an art. Groucho Marx called it "the art of looking for trouble," and Will Rogers said it was "applesauce." Whatever it may be, politics clearly doesn't lend itself to engineering analysis in the traditional sense, as engineers should have learned by now.

## THE PERSNICKETY PROFESSION

Lack of political savvy is only one of several problems that engineers face when they venture out into the world. I fear that in general they—we—suffer from a tendency to be persnickety, that is, overly fussy. It may seem paradoxical to say that engineers are often meticulous to a fault, but I have repeatedly seen this to be so.

When I was a student in engineering school, I assumed I was part of a fellowship dedicated to good works. The engineering profession seemed all of a piece, an association of talented people who sought solutions to technical problems. Occasionally I argued with my fellow students about which design was most elegant or economical, but invariably—counseled by a professor—we ended in agreement. It seemed that arriving at the "best" answer was merely a matter of competence and experience. It did not occur to me that in the "real world" the preferred solution to a technical problem might be subject to debate between individuals with differing social objectives, much less that engineers themselves could clash in bitter disputes.

It was only after I left academia that I discovered how splintered the engineering profession is, and how dramatically engineering judgment can be affected by personal idiosyncrasies, by one's role in the community, or simply by one's job. To some extent this is good: Out of wholesome disagreement come benefits for society. But the danger is always present that engineers will (1) be tempted to cut corners when it is to their economic advantage; or (2) be excessively exacting when they have no responsibility for the costs entailed. The risks of (1) are fairly obvious, and much care is taken to protect against them. But not nearly enough attention is paid to (2), which I have found to be a serious and costly defect in the engineering process.

As a civil engineer working in the construction industry, I have encountered designers whose work entails unnecessary expense. The design engineer is a sort of god—the project specifications are sometimes referred to as "the bible"—and doubtless it is tempting to specify, regardless of cost, the most exacting methods and the most expensive equipment. Engineers who follow such a course, heedless of their clients' best interest, are usually winnowed out by the exigencies of the marketplace. More subtle—and more insidious—is the influence of inspectors, those engineers who are charged with verifying that contractors comply with plans and specifications. It is absolutely essential that these inspector-engineers be totally honest and vigilant. Quality control depends upon them, to say nothing of public safety. But it is also important for inspectors to be *reasonable* in interpreting contract documents. Since in erecting a building there is no such thing as perfection, what deviation from the ideal is permissible? How level is level? How smooth is smooth? Tolerances can be spelled out numerically; but there invariably comes a point where common sense, traditional practice, and pragmatism enter the picture.

My feelings about this are forever affected by an experience I had on one of my very first jobs. The incident was so bizarre that it would have been laughable, except that it was utterly serious and involved a potentially large loss for my employer.

We were putting the finishing touches on a five-story masonry building when a project inspector summarily ordered us to remove the entire front facade. He judged some of the bricks to be excessively dark in color, a sign that they had been overburned in the kiln. The architect, who rules on matters of aesthetics, was willing to accept the contrast, but the engineer-inspector remained adamant. Not only did he consider the color range unacceptable, but he feared that the overburning had affected the strength and porosity of the brick, making it liable to disintegrate or absorb water.

The ensuing imbroglio involved many additional personalities: the engineer who had designed the structure under contract with the architect; the chemical engineer who worked for the brick manufacturer, and another who represented the brick manufacturers' trade association; the engineers who worked for the testing laboratories brought in by both sides of the dispute; and by implication, although not in person, the many engineers who had developed the standards that were cited in the contract specifications.

Eventually all parties concluded that the installation was safe, durable, and watertight. Nevertheless, the issue of permissible variation remained unresolved. The inspector argued that if the brick were accepted the contractor would have "gotten away" with something. At the very least he demanded that some penalty be assessed. The matter was referred to the Owner.

As luck would have it, the Owner in this case was the Catholic Archdiocese of New York, and the person designated to rule on our appeal was a priest. I will never forget the occasion. The hearing took place in a splendidly decorated ecclesiastical chamber, and the decision was rendered in the form of a sermon. The lesson of the day—preached through homilies and biblical references—was that in the nature of things bricks should be expected to come out of the kiln with slight variations, and that the inspector was looking for an unreasonable degree of precision, a perfection to be found only in heaven.

That worthy cleric has long since gone to his reward. But through the years I have repeated his words, with a variety of embellishments, to many an inspector—not always, I must confess, with the desired effect. Wherever contractor-engineers cross swords with inspector-engineers, there are inevitably possibilities for tragicomedy.

Another experience in the absurd occurred more recently when my company was completing a residence for the indigent elderly. We were in the process of obtaining a certificate of occupancy when a mechanical inspector for the supervising government agency informed us that the Community Room did not have adequate ventilation. The building code required that the area of the window openings in each room be at least 5 percent of the floor area of the room, and the inspector ruled that we were slightly short of compliance. It seems that the windows everyone else had assumed were 8′-0″ square were only 7′-8″ square when the *operable* sash were measured (the frames being two inches wide). The consulting engineer who had designed the building's mechanical systems reviewed his calculations and concluded that, considering the fan-forced ventilation in the room, there was more than adequate fresh air. But the inspector stood firm on his literal interpretation of the code. Even his own superiors, arguing on behalf of common sense, could not dissuade him. The scheduled formal opening of the project was only days away; calamity loomed.

Meeting followed meeting as a procession of engineers—inspectors, consultants, contractors, building department officials, and eventually administrators and commissioners—grappled with the problem. As tempers grew short, I thought back to the naive assumptions of my days in engineering school. There was no professor to give us the correct solution to this problem.

Happily, a scholar of sorts did arrive in time to save the day. The inspector's superior was an old-timer, an engineer wise in the lore of the building code. He informed us that all would be well if we would post a sign indicating that when the room was crowded, occupancy was limited to two hours! And that is what we did, after filing an appropriate amendment with the Building Department.

The opening ceremonies took place on schedule—in the very Community Room that had nearly been our downfall. With the mayor of New York and other politicians on hand, a band playing, and the room filled to capacity, I must confess to having fleeting concerns about the ventilation. But I need not have worried. The air remained fresh and everybody smiled—including the engineers who had recently been at each other's throats.

## A LACK OF PRIDE

I have said that engineers tend to be persnickety; but in one particular respect they are disturbingly casual. I refer to the matter of professional engineering licenses, which most—more than 80 percent, in fact—don't bother to obtain.[5] I think that this is a mistake, partly for practical reasons, but also because this indifference shows a lack of professional pride. My misgivings were brought into focus as a result of two singular encounters that I had in quick succession a number of years ago.

One cold winter's day, high in the superstructure of a building under construction, I joined a group of men who were peering at the dial of a deflection gauge. Because of doubts about the strength of a concrete slab, a twenty-four-hour load test had been imposed, and several engineers—representing the Building Department, the structural designer, the testing lab and the contractors—were gathered to record the results. A glance at the dial showed that the loaded slab had not deflected beyond allowable limits, a happy result for those of us who would bear responsibility for any required remedial work. But the chief inspector from the Building Department was not about to let the matter end without appropriate formality. He took out a long legalistic-looking document and began to fill it in: date, time, location, temperature, and names of all those present. He then asked each of us for the registration number on our state professional engineer's license. After much fumbling about in our wallets we produced the requested information, and he recorded it carefully on his form. Aside from a few jocular comments about whose numbers were the lowest, and how this reflected upon our comparative ages and experience, he treated the moment with great solemnity. Clearly he viewed the professional license as a meaningful credential and the holders of this license as members of a select company. I found the occasion curiously moving.

Some days later, while visiting an IBM facility in North Carolina, I was surprised to hear several engineer-executives disparage the concept of state licensing. They conceded that licensing serves a purpose for consulting engineers who have to file plans with municipal building departments. But for creative sorts like themselves, and for the vast majority of engineers who work in large corporations, they considered the notion superfluous, undemocratic, and pretentious. Besides, they added—to make sure that I did not misunderstand the nature of their objection—the PE examinations are ridiculously easy, not at all a test of professional competence. This discussion stood in vivid contrast to the experience I had shared with the PEs gathered around the deflection gauge.

Anyone who has considered, however superficially, the topic of licensing for engineers, is familiar with the dilemma. Indeed, I should have known better than to expect a different reaction at IBM. The vast majority of American engineers *do* work for large corporations and will never have occasion to file plans with a government agency or need to "place their seal" on their work product. Because they do not need a license to earn an accredited degree from an engineering school, to get a job as an engineer, or to move successfully along a career path, they show profound disinterest in the matter.

In order to obtain a professional license—the process has been integrated among practically all the states—a graduate of an accredited engineering school must first take the Fundamentals of Engineering (FE) examination, also known as the EIT (for the Engineer-in-Training certificate awarded to those who pass). Then, after four years of practice, one qualifies for the final exam, which is less technical than the FE and stresses general concepts of professionalism including ethics and the making of sound economic choices.

The ideal time to take the FE exam is just before or immediately after graduation from engineering school, when the technical material is freshest in mind. Yet surveys show that scarcely more than 40 percent of graduates are doing this. Only 13 percent of engineering schools require their students to take the exam, and only 3 percent require them to pass.[6]

Regardless of what esteemed engineers in large corporations may think or say, I consider this situation deplorable. Becoming a professional in today's world should certainly include obtaining a government-sanctioned license. Membership in a voluntary professional society is no substitute for licensing, not least because in recent years the courts have sharply restricted the power of these societies to discipline their members. It is also a sad truth that fewer than half of American engineers are members of such societies.

In recent years we have heard a lot of well-intentioned platitudes about engineering ethics from both inside and outside the profession. What could better manifest our seriousness in this field than to require all engineers (or at least future engineers) to obtain licenses? It is amazing what having a license—something that has been hard-earned yet can be taken away—will do for the conscience. I have heard many a consulting engineer say, "Why, I couldn't think of doing that; it could cost me my license!" This notion might find a place in industry just as it has among consulting engineers.

Beyond issues of ethics and discipline, however, there lies the question of pride in profession. Without a serious respect for licensing, engineering sets itself aside from its fellow professions, not only medicine and law, but accounting, nursing, and teaching, among others. By what we do—or fail to do—we make a statement about our moral priorities.

It may well be true that no test can establish competence nearly as well as the earning of a degree from an accredited engineering school. And admittedly a fondness for credentials and ceremony can lead to shallow bombast. Still, engineering is more than just a job; it is a profession, and in some respects a "calling." Engineers are committed to serve the community. Acceptance of state licensing makes this commitment manifest.

### FOR WANT OF A LAUGH

Of all the flaws that might be found in the typical engineer, the one that I find most vexing relates to general disposition, to a certain lack of lightheartedness. It is all very well to be serious; life is not a bowl of cherries. But there are times when the lack of a sense of humor is unhealthy and can damage good causes. For example:

A while ago Russell Baker, the well-known syndicated columnist, wrote of the complex new phone system that had been installed in his office. He confessed that he was afraid to take lessons in how to use it. What if he flunked the course? "It would be humiliating, having to go to summer school to make up a failure in Telephone."

He went on to express the bemusement of the average person trying to cope with rapidly changing technology. Although the telephone originally "was a swell idea," he finds several of its latest developments—for example, voice mail—both baffling and bothersome. The droll essay was entitled, "March of the Engineers."[7]

A few weeks later, in another essay, Mr. Baker reported that several engineers had written to him to complain. One correspondent had accused him of "a narrow-minded and pernicious antitechnology attitude." "Worse," noted the columnist, "he thinks I contribute to the decline in the number of Americans studying engineering." After assuring his reading public that he appreciated the importance of technology, and engineering education in particular, Mr. Baker gently went on to chide his engineer critics for becoming so indignant. In a parting shot, he restated—humorously yet earnestly—his plea for user-friendly telephones, and for a VCR that might be programmed "without a degree in electronics."[8]

Is it possible that we engineers don't have a sense of humor? It's no secret that as a group we are not jolly, light-hearted, or much given to jokes. Occasionally one hears of technically ingenious pranks, such as when engineering students take apart a car and rebuild it in a dormitory room, or modify a classroom clock so that it can be sped up or slowed down to fluster a professor. But most engineers are sober and serious, particularly when discussing their work. According to one survey, three quarters of American journalists agreed that engineers could aptly be described as "wooden."[9] I think that's a bit harsh. I meet a lot of engineers, and generally find them congenial company. But engineers do not suffer fools gladly, and I fear that they are not as tolerant as they might be when it comes to determining who is a fool.

The trouble doubtless begins with the precision inherent in engineering work. Technical problems usually have correct answers, or at least optimal solutions. Engineers learn to be suspicious of whimsy, caprice, and absurdity, the very stuff of humor, but dangerous notions when public safety is a consideration. The situation is made worse by the people who malign engineers—the antitechnologists—many of whom are themselves humorless and bellicose, and a few of whom are patently foolish to boot. So, influenced by the nature of their own professional work, and goaded by their critics, engineers become irascible and end up scolding an amiable newspaper columnist.

To the extent that engineers have indeed lost their sense of humor, the profession and society at large pay a heavy price. As Russell Baker noted, many folks have problems coping with changing technology, and if engineers are austere and aloof this situation is exacerbated. Equally important, if engineers are to participate in the great communal debates, and take on leadership responsibilities, as I think they should, then they must learn the niceties of discourse, beginning with wit and empathy. If one wants to be persuasive, grumpiness is not the key. I favor humor for serious reasons as well as for love of a joke.

A notable exception to the sober if not angry tone adopted by so many engineers is the bimonthly "Reflections" column written for many years by Robert W. Lucky in *IEEE Spectrum*. As executive director of research at AT&T Bell Laboratories, and then vice president of Bellcore, Lucky has stood at the forefront of the engineering community, which makes his merry spirit all the more precious. He has directed his humorous barbs at such likely targets as the guilt of having fun at technical conferences, the inevitable failure of equipment demonstrations in the presence of Members of Upper Management ("the MUM effect"), and communication within organizations ("We all have the comforting illusion that there is someone up there who knows what is going on."). But he has also been able to apply a light touch to such topics as citizen advice to government, the public image of engineers, and the loss of American preeminence in certain technologies.

Bob Lucky's columns are cheering not only in themselves, but also because they appear in the official publication of the largest engineering society in the world. One can only infer that some engineers resonate to his spirit. May their numbers multiply. We need to cultivate humor in the engineering community. This is no laughing matter.

Ineptitude in politics, a tendency to be persnickety, lack of professional pride, and occasionally the absence of a sense of humor: These cannot be called crippling flaws of character. Nevertheless, to the extent that they are surmounted, the cause of engineering will benefit.

There is reason to be hopeful. Engineers are trained to be observant and to learn from experience. We will learn from our mistakes; trial and error has always been a key element of the engineering method.

CHAPTER 9

# DOING GOOD

## UNWILLING TECHNOCRATS

Most engineers are willing to admit that they have faults and foibles. But to critics who call them "today's technological aristocracy"[1] or "our new masters,"[2] engineers can only stare with disbelief. In spite of all evidence to the contrary, a number of nontechnical people—some of them skilled in the use of words—continue to warn against the domination of social process by so-called technocrats.

In *Blaming Technology,* a book I wrote in 1981, I tried to demonstrate that engineers as a group are not a power elite with authority to dictate the directions in which society moves. We do not have the money or the social cachet. We have little say in the worlds of finance, politics, academe, and the media. The technologies we create survive only if they meet the requirements of the public as demonstrated in the marketplace. Only one technology out of a hundred chosen by a leading think tank for development survives as a viable product.[3] The statistics that I brought to bear in 1981 have not changed much if at all in the intervening years. It is a great cop-out for intellectuals or anyone else to maintain that engineers as a group are to blame for societal problems.

Along with unjustified censure, engineers are burdened with irrational expectations. Some of the engineer's severest critics have concluded that the profession might transform itself and become a savior of society. This is, to say the least, ironical. If these critics-turned-disciples looked to engineers for creativity and potential leadership, the response would doubtless be along the lines of thank-you, we will do our best. But when they call upon engineers to assume the role of policeman and guardian of industrial morality, that is quite a different matter.

Some few years ago I attended a seminar sponsored by the National Research Council's Committee on Education and Utilization of the Engineer. A number of historians had been invited to participate in the three-day conclave, and in their formal papers they examined contributions made by engineers to the development and well-being of the United States. They also raised pertinent questions about the evolving role and status of engineers in American society. Several speakers contrasted the illustrious "chief engineer" of a century ago with the typical engineer of today, who works for a large corporation and has very little independent power. In some of the presentations, this historical fact was expressed in terms of regret, and as an engineer I shared that regret. In terms of engineering prestige, the good old days sounded very good.

A few of the speakers, however, went beyond bittersweet comparisons. Adopting a missionary tone, they called upon today's engineers to recapture the power of their professional forebears. They suggested that engineers could accomplish this by refusing to yield the decision-making role to corporate executives. Industry, it appeared, was an antisocial villain; engineers were society's potential saviors, if only they would stand up and be counted. Specifically, engineers were called upon to "have a sense of their own independence," and further, to "claim engineering authority."

On the final day of the conference, when one of the rapporteurs failed to appear, I was asked to attempt a summation. In doing this, I took issue with the view that I had heard expressed. I pointed out that conditions had changed since the time when robber barons ruled American industry, when there were few laws or regulations to protect public safety, and when an honest engineer had to play the role of "a cowboy in a white hat," as one speaker had put it.

In a lawless society the gun-toting cowboy may be a hero; but once law has been established he becomes a vigilante. It is not rational to tell contemporary engineers within industry that they should attempt to wrest power away from executives. Engineers themselves are often executives. "If historians don't like the structure of corporate America," I said—perhaps more flamboyantly than was prudent—"they shouldn't expect engineers to start a revolution. Let them start their own revolution."

In the fall of the following year, I was invited to attend the annual meeting of the Society for the History of Technology, and asked to respond to a panel who addressed the topic, "Defining Engineering in the Twentieth Century." Again it seemed to me that several of the historians voiced unreasonable expectations concerning engineers, particularly electrical engineers who participate in the development of armaments. And again I responded with perhaps more fervor than tact.

Of course I do not believe that these academics are literally trying to foment revolution by politicizing engineers. (Thorstein Veblen did advocate this around the time of World War I, but he didn't get very far with it, and during the Depression years the "technocracy" movement came to be generally discredited.) Nevertheless, there *are* some historians and social critics who seriously espouse the view that engineers can and should attack social problems by exerting "professional" influence within the workplace, by standing up against the pressures supposedly exerted upon them by "businessmen." These people believe that our difficulties with product safety, environmental pollution, and armaments—just to name three of the most troubling technological problems of the age—could be ameliorated if engineers would only act in accordance with appropriate ethical standards. After all, they argue, this is how engineers used to behave before they "sold out" to industry.

A prominent exponent of this view is Edwin T. Layton Jr. a respected historian of technology, whose 1971 book, *The Revolt of the Engineers,* has become a classic in the field. It has indeed become a bible for those who would hark back to days when engineers purportedly tried to express their social conscience. In his preface to the 1986 reissue of this work, Layton called upon engineers to form a "loyal opposition" within corporations. In this he echoed the theme of his 1983 essay, "Engineering Needs a Loyal Opposition."[4] He said much the same in his introduction to *The Making of a Profession,* the centennial history of the Institute for Electrical and Electronics Engineers published in 1984. In that introduction he praised the volume's author, A. Michal McMahon, for calling attention to "the new professionalism"—a heightened sense of social responsibility that he perceived growing among electrical engineers, particularly after the Vietnam War.

We are all in favor of morality, and one can only respect Layton, McMahon, and their like-minded colleagues for their worthy intentions. But the trouble is that in complicated technological matters, the definition of "safe" or "appropriate"—the recognition of "right" or "good"—is often subject to debate between individuals, even very well schooled individuals. Recognizing this, our society has gradually created democratic ways of deciding how to cope with such problems. There are today thousands of laws, regulations, and codes that give guidance where just a few years ago there was no guidance. And where specific regulations do not apply, concepts of legal liability have evolved that supersede previously vague concepts of moral responsibility.

For example, tort law used to require that negligence be proved as a "proximate" cause of injury, but since the 1960s, courts have been imposing "strict liability" for injuries *whether or not* negligence is involved. As stated by the authors of *Product Safety and Liability: A Desk Reference,* "The one choice engineers, designers, and their employers no longer have is whether or not to pursue safety goals. To ignore or pay only lip service to safety . . . can jeopardize corporate survival."[5]

Thus the physical well-being of the public no longer depends upon the heroic actions of individual engineers. This is not to say that dishonesty and corruption have disappeared, or that there is no place for the occasional valiant whistleblower. But the exception—the very rare exception—should not be raised as the keystone of engineering ethics.

People with special knowledge and skills do have special obligations to society. This is basic to our communal understanding of professionalism. But it is important that professional decisions not be prejudiced inappropriately by individual politics. We don't want engineers bringing personal whims to bear when making technical decisions.

The academics and ethicists who favor a "new professionalism" and the concept of "loyal opposition" seem to assume that engineers of conscience will be motivated by a sort of kindly progressivism. On the other hand, at least one professor of philosophy, writing in the *Business & Professional Ethics Journal,* is not so sanguine. Paul T. Durbin of the Center for Science and Culture at the University of Delaware observes that: "The overwhelming majority of engineers I have gotten to know well are personally and often politically conservative. They are enthusiastic believers in the corporate philosophy and vociferous opponents of government regulation."[6] I do not maintain that engineers are mainly left or right politically, nor do I know of any reliable studies on the topic. Whatever the case, I am most comfortable—as a citizen, and also as an engineer—to have professionals bound by democratic processes. (I wonder how many teachers of engineering ethics have read Ayn Rand's novel, *Atlas Shrugged,* in which the nation's professional engineers go on strike against "liberalism"?)

I do *not* mean that we should be nonchalant about our professional responsibilities or not continue to think seriously about the meaning of engineering morality. However, rather than mouthing vague or outdated platitudes, we must face up to the fact that the role of the engineer in society is not—and should not be—the same as it was a generation and more ago.

There is no shortage of ways in which engineers can express devotion to noble professional concepts. As a start, it is imperative that engineers be conscientious in their work. This may seem trivial until we stop to think that many of the worst technological disasters, such as Bhopal and Chernobyl, stem mostly from carelessness and ineptitude. It is also important that engineers do pro bono work and take responsibility for educating the public in technological matters. I admire those engineers who work in the nonprofit sector: teaching, doing research, and overseeing public works. Further, I endorse the banding together of engineers of like mind, such as those who signed a pledge not to work on Star Wars—or, with equal democratic right—those who might have wished to endorse Star Wars.

As for the professional societies, there are many issues on which they have an obligation to take public positions and appropriate action, for example, fostering support for research and science education, and sponsoring studies on the health effects of certain technologies. But when it comes to politics and controversial moral issues, there it seems only proper that the societies should encourage tolerance and debate. How can one help but be impressed by IEEE polls that show the membership split practically down the middle on the question of whether one can be proud to work in the field of weaponry? Indeed, almost any poll of engineers on any topic—including issues of engineering ethics—will reveal a great diversity of opinion.

In the end, one's stance on any issue is based upon personal morality. Naturally, one may endorse a shared standard of virtue, whether religious or otherwise. But we dare not define "good" engineering by the precepts of any particular religious, philosophical, or political group. We may jointly pledge to work on behalf of the "public welfare," but our ideas about exactly what that welfare is, and how it is to be pursued, are mostly individual, not profession-wide. The well-intentioned critics who would have engineers become vigilantes have not thought the issue through clearly.

It is troubling to debate this topic, as I do from time to time, and to argue against well-meaning idealists. But the first duty of an engineer is to be realistic. I will have more to say about idealism in the final chapter of this book.

## WHISTLEBLOWING: INSTITUTIONALIZED MARTYRDOM

Let me try to be as clear as I can be: By disavowing the role of "loyal opposition" I do not seek to evade the concept of professional responsibility. The engineer is obliged to protect the public interest, and on occasion—much more rarely than the teachers of engineering ethics would have us think—this may require the drastic act that we have come to call "blowing the whistle."

A vexatious business. In concept a simple act of conscience, but in life fraught with awesome complexities.

Probably the most renowned engineer whistleblower of our time is Roger Boisjoly, the chief engineer at Morton Thiokol, who before the *Challenger* disaster repeatedly warned his superiors about inadequacies in the joints of the shuttle booster. The night before the ill-fated launch, he led a group of Thiokol engineers who argued urgently but vainly against proceeding. After the shuttle disaster, he was candid and outspoken at government hearings, revealing all that had transpired before the launch and during the crucial decision-making process. I followed with fascination and dismay the events that ensued.

As might have been predicted, Boisjoly's bosses and many of his colleagues treated him as a pariah. Six months after the accident, no longer able to endure what he termed the "hostile" environment at Morton Thiokol, he left the company on extended sick leave. Eventually he was diagnosed as having a post-traumatic stress disorder and underwent two years of psychological therapy.

After recovering his health, Boisjoly embarked on a series of lectures that he said gave him a "positive catharsis." In speeches at several dozen universities, and addresses before most of the major professional engineering societies, he condemned corporations that "make arbitrary and irresponsible decisions that kill people and ruin the lives and careers of their employees without accountability."[7] He urged engineers to stress moral responsibility and professional ethics.

Boisjoly's personal misfortune—publicized by his many lectures and interviews—lent force to the growing crusade on behalf of whistleblowers. Proponents of this movement urge stronger laws to protect employees who speak out, and call upon professional societies to lend them increased support. It has even been suggested, in a *New York Times* op-ed essay, that we honor prominent whistleblowers by putting their pictures on postage stamps.[8]

I found that the Boisjoly tragedy, rather than renewing my enthusiasm for the cause of whistleblowing, started me thinking in completely different terms. Although I support the laws and processes intended to protect the individual who speaks out against his organization, I am distressed by the anguish that invariably ensues. We may admire whistleblowers for their courage and concern for the public good, but there is no escaping the fact that what they do usually entails an element of personal disloyalty. We may protect whistleblowers economically, and even give them awards, but nobody can protect them from the hostility of colleagues who feel betrayed. Enduring this hostility—which often means tacit blacklisting within their industry—becomes a form of martyrdom.

Why do we not exert our efforts toward putting in place systems and procedures that will eliminate—or at least greatly reduce—the need for such martyrdom?

NASA made strides in this direction when they turned anew to the shuttle program after the *Challenger* disaster. *Restructuring* became the watchword, and actions followed hard upon proclaimed intentions. The agency created a new post: Associate Administrator for Safety, Reliability, Maintainability and Quality Assurance, an office reporting directly to NASA's Administrator and having the authority to stop shuttle launches. They announced that the size of the safety and quality staff would be increased by more than a third and would be trained to be more aggressive in ferreting out problems and more vocal in demanding that they be corrected. (Since 1970, while the number of NASA employees declined 31 percent, quality control personnel had been reduced by 70 percent.) The agency started to do more of its own analyses of part failures, supplementing the analyses done by contractors, and paying particular attention to analysis of trends in defects reported after shuttle flights. The lack of such mandatory trend analysis was one reason that Boisjoly's information about the rocket joints' weakness in cold weather was not appropriately disseminated. Other organizational changes were intended to clarify lines of authority and to eliminate rivalries between the Marshall and Johnson centers, rivalries that contributed to lack of communication on crucial issues, including the rocket joint problems.

The improvements that can be achieved through organizational change should not be underestimated. For example, in New York

City, building collapses and other serious accidents were habitually attributed to careless or unscrupulous builders and building department inspectors. After decades of bitter experience, and fruitless demands for caution and honesty, a new system was put in place. Today, before one can obtain a certificate of occupancy for a building, independent licensed engineers must certify to the integrity of concrete (both as mixed in the plant and as poured in the field), all welds and other connections, firestopping (walls, floors, and ceilings designated to stop the spread of fire), sprinklers, and about two dozen other safety features. This process is bothersome, costly, and time-consuming. But it works.

Speaking of licensed engineers, Boisjoly has speculated that the *Challenger* disaster might have been averted if he and his colleagues had been licensed professional engineers. (As noted in the preceding chapter, most engineers who work for large corporations do not bother to obtain licenses.) The warnings of licensed professionals, says Boisjoly, would carry additional weight with the ultimate decision makers. Certainly this is a cause worth promoting, as I have already argued.

Many corporations have restructured their safety operations, motivated at least in part by new laws and regulations, and also by the judges and juries who assess liability even in the absence of proven negligence. An industrial firm can be held liable for defective product design, defective manufacture, inadequate labeling, and faulty packaging. Companies can be penalized for failing to keep proper records of product sales and distribution, project failures, and customer complaints. Consequently, appropriate changes are introduced into the routine of doing business. Such operational changes afford better protection for the community than an army of whistleblowers.

In *Fortune,* an article entitled "Listen to Your Whistleblower" advises managers to use an ombudsman system, or at least to cultivate an extensive personal "grapevine" among their employees.[9] Many companies have long had "open door" policies. IBM, for example, guarantees the right of any employee to appeal a supervisor's decision without fear of retaliation. In recent years about two hundred major companies have set up formal systems in which a senior executive outside the regular chain of command is available on a "hot line" to deal with employee grievances and alarms on a confidential basis.

We will never eliminate the need for honorable and courageous technologists, nor would we wish to. But a system that relies upon heroism is neither stable nor efficient. A society that expects martyrdom from its citizens is neither wise nor noble.

Perhaps Roger Boisjoly's greatest contribution will lie not as much in encouraging others to emulate his actions as in rousing the rest of us to develop systems and procedures that will make as rare as possible the occasion for sacrifices such as his.

## GRIME AND PUNISHMENT

In spite of best intentions, and even after we have implemented the best safety systems we can devise, bad things will happen. Sometimes the more we moralize the more we frustrate the very actions that the situation demands. According to the cliché, actions speak louder than words. Nowhere is this more true than in the field of technological oversight and regulation.

For example, I see by the morning paper that a terrible odor, said to be like "rotten eggs in a sewer," is afflicting the residents of a Long Island community with the sadly ironic name of Flower Hill. The source of this stench is the nearby landfill of the town of North Hempstead. It appears that for several months hundreds of truckloads of garbage and construction debris—some from faraway sources in New York City—were illegally dumped in an unlined section of the landfill property. Nobody seems to know how this was allowed to happen. The man who headed the Sanitation Department when it occurred has recently died, and other town officials disclaim responsibility. The toxics coordinator for a local public interest group indignantly points out "how easy it is to create problems when the D.E.C. doesn't enforce things right and the solid waste officials are lax in their efforts."[10] Indeed.

Happily, the Flower Hill garbage is not toxic and the hydrogen sulfide it produces, while obnoxious, is not a health hazard. Much worse is the case of the landfill in Port Ivory, Staten Island, where thousands of tons of asbestos and infectious waste from hospitals were surreptitiously mixed in with the "clean" construction debris that the site was licensed to receive. The responsible parties were prosecuted and found guilty of racketeering. They agreed to pay $22 million in penalties, and several were given lengthy jail sentences. (As in the Flower Hill chronicle, one of the principal suspects died. The circumstances of his demise were more noteworthy: He was shot to death while out on bail after the conclusion of the trial.)

Tales of environmental crime are increasingly in the news. They attract the attention of the press in part because of the derring-do involved. The furtive disposal of poisons becomes the contemporary equivalent of a stagecoach heist, with the culprits seen as audacious desperadoes. Certainly this was the case when front-page stories recounted the deeds of Evelyn Berman Frank, the seventy-five-year-old "Dragon Lady" of New York Harbor. As matriarch of family-owned companies that hauled sludge and petroleum products, and also cleaned tanks and ship hulls, Mrs. Frank had for twenty years flouted the law with repeated incidents of dumping sludge, acids, and other pollutants. This saga came to public notice when one of her barges sank in the harbor, spilling forty thousand gallons of oil, and Coast Guard officials found that it did not have a valid permit. In fact, the vessel had been denied a certificate of inspection because the Coast Guard considered its hull too weak to carry oil safely. Investigation revealed that a number of interlocking corporations controlled by the Frank family owed more than a million dollars in New York, New Jersey, and federal fines, taxes, and judgments. In a much-publicized denouement, additional fines were levied and paid, and Mrs. Frank was sentenced to five years probation. Had she been younger, said the judge, she would have gone to jail.

For all the titillation provided by these tales of villainy, the sobering fact is that criminal pollution has become an extreme problem, and there is every reason to believe that it will get worse. It is all very well for ethicists to expound on the need for engineers to protect the public interest. It is admirable to see government passing laws intended to help purify our air, water, and soil.

But to what avail are noble intentions, and even exemplary legislation, if we lack the strength of will to turn words into reality? Ironically, the stricter we make our antipollution regulations, the more alluring we make the enterprise of waste disposal to criminals. Most industrial corporations will abide by the law, however grudgingly. But racketeers move in precisely where illicit profits are the most lucrative.

Once we recognize the gravity of the problem, and once we acknowledge that the solution is not to be found in a mere accumulation of well-intentioned laws—much less in moral platitudes—what do we do next? Obviously, we should seek through political pressure to obtain funding for government regulatory agencies so that they can hire more inspectors and otherwise enforce the regulations they are charged to administer. It is absurd to pass new laws while cutting back on the budgetary resources needed to make these laws effective. Yet this defect in American democracy long predates the environmental movement.

There are some hopeful signs. A National Environmental Crimes Prosecution Center has been established in Alexandria, Virginia. Funded by the Justice Department and the National District Attorneys' Association, it serves as an information clearinghouse. Many states now have special environmental-offense units within the offices of the attorney general. Several states have created trained squads in each county. Such units investigate environmental crimes much like other felonies. Informers are used in about half of the cases, and environmental forensics is a rapidly growing field. Still, the resources are not equal to the need.

Good things are happening in the voluntary sector, but here, too, funds and personnel are meager considering the scope of the problem. In North Bergen, New Jersey, members of a town-sanctioned citizen "posse" patrol their community issuing summonses to illegal dumpers. On a national scale the Izaak Walton League has more than three thousand streamwatchers demanding that regulatory agencies act when water pollution is observed. Boats funded by public interest organizations patrol the Hudson River, Long Island Sound, the Delaware River, Puget Sound, and New York Harbor. A number of laws such as the Clean Water Act provide for paying the legal costs of watchdog groups who bring successful suits against violators. Worthy as such efforts are, one has to question how citizen groups will fare as organized crime moves more intently into the waste disposal enterprise. About as well, one might suppose, as they do in coping with the spread of illegal drugs.

In this connection, it is worth noting that Congress has allocated funds to the armed forces to assist in drug interdiction operations. AWAC airplanes, outfitted with sophisticated radar equipment, have been used to fly over the Caribbean, and a network of radar-carrying balloons was for a time deployed across the southern United States. There has also been a plan to have National Guard units, with drug-sniffing dogs, search cargo shipments in American ports. Is it unreasonable to suggest that the Pentagon might be enrolled in the emerging war against illegal pollution?

While political approaches to the problem are being debated, we should commission engineers to work on technical methods of detecting and foiling the outlaws who, for profit, seek to contaminate the environment we are increasingly learning to value. Another engineering strategy lies in devising ways of making acceptable disposal (as well as pollution cleanup) less costly, thus making the enterprise less attractive to criminals.

In approaching the problem of protecting public health and safety, as in so many other areas, Americans fail to heed the lessons of history. As George Washington once wrote, "We have probably had too good an opinion of human nature . . . Experience has taught us that men will not adopt and carry into execution measures the best calculated for their own good, without the intervention of a coercive power."[11]

Our community is perpetually under siege. Factories explode, buildings collapse, trains derail, criminals plunder and kill, and, most ominous of all, terrorists threaten unthinkable cataclysms. We must constantly be on guard—against stupidity and carelessness, greed and rapacity, hostility and fanaticism. We need good engineers: talented, imaginative, and committed. We want them to have high ethical standards. But we do not help ourselves by being unrealistic and mouthing clichés.

CHAPTER 10

# LEADERSHIP

## THE TIME IS AT HAND

I have argued against turning engineers into vigilantes, solitary guardians of the good. Yet, I believe with all my heart that there is a compelling need for engineers in high councils. With politicians in ever greater disrepute, and technological issues daily more portentous, the case seems self-evident. Of course, the idea of using "experts" to help run governments is as old as centralized government itself, and the lessons of history are far from reassuring.

In the fourteenth century, for example, a sultan of the Ottoman Empire began gathering outsiders—mainly Christian youths captured in war—and, after converting them to Islam, trained them under the strictest discipline to become loyal servants of the state. Thus were formed the Janissaries, a corps of soldier-bureaucrats who served effectively as the Ottoman Turks came to dominate much of the Eastern world and threaten the security of Europe. However, the power vested in this select group was often used to intimidate the state rather than serve it, and by 1600, corruption became prevalent as membership was achieved largely through bribery and then became hereditary. In 1826, Sultan Mahmud II rid himself of the Janissaries by the simple expedient of having them murdered in their barracks.

History teaches that no single group, whatever its training and talents, can "run" a society satisfactorily, at least not for long. Yet, the idea that political power belongs in the hands of trained professionals has always had adherents. Plato, in his *Republic,* proposed that the ideal society be governed by an assembly of philosophers. The concept has never actually been tried, which is probably just as well. In the United States, during the desperate days of the Great Depression, a few intellectuals—and more than a few eccentrics—called for engineers to take charge; but, as I have noted earlier, the short-lived fiasco of the "technocracy" movement helped to discredit this notion.[1]

Nevertheless, within the guidelines of our traditional democracy, the time is at hand for engineers to take a more active role in conducting society's affairs. More than ever we require leaders who understand engineering, and engineers who are capable of being leaders. This is the "pull" side of an evolving phenomenon—the demand that will arise—must arise—out of the radically changing economic and political situation.

There is also a "push" at work—a welling up of desire within the engineering profession—a yearning for something better and more noble. There are many factors that contribute to this aspiration. One relates to our strong feelings for the future. If we are crucial agents of change, if technology is to bring new prospects of hope for humankind, then it seems necessary and right that we should take on new roles of leadership. Paradoxically we are also in love with the past, a partly mythical past perhaps, in which engineers were heroic figures in the community, and we want to restore elements of that vanished golden age.

I believe that this new awakening of ambition is in some ways intensified by the discrediting of Communism. We sense the possibilities of a better world, a renaissance of sorts, in which the elegance of the past will be wedded to the democratic ideals that have prevailed. Aleksandr Solzhenitsyn has described how engineering changed during and after the Russian Revolution. He didn't like the engineers who were educated under the Soviet regime. He found them stern and impersonal, inferior to the older generation both "in the breadth of their technical education" and "in their artistic sensitivity and love for their work." Of the older engineers, many of whom he met in the *gulag* prison camps, he wrote with glowing affection. He admired "their open, shining intellect, their free and gentle humor, their agility and breadth of thought, the ease with which they shifted from one engineering field to another, and, for that matter, from technology to social concerns and art." "Then, too," he continued, "they personified good manners and delicacy of taste; well-bred speech that flowed evenly and was free of uncultured words; one of them might play a musical instrument, another dabble in painting; and their faces always bore a spiritual imprint."[2]

I do not suppose that such engineers will come again—if indeed they ever existed as Solzhenitsyn depicts them. Yet we can't help being captivated by the ideal. Part of what we want to be lies hidden in something that we, as a profession, used to be and are no more—not only in Russia, but wherever insensitive materialism has taken hold.

The Russian example is not entirely helpful to my argument, of course: The majority of Soviet leaders were educated as engineers, and see where that got them! Yet, according to physicist Sergei P. Kapitza, who left the Russian Academy of Sciences to become Visiting Fellow at Cambridge University, the Soviets debased their engineering educational system, seeking technicians instead of true professionals, with disastrous results. "In fact," writes Kapitza, "the Soviet Union never really managed to produce an engineering elite, an upper class of highly placed, well-paid and respected engineers, which is the backbone of any modern industrial nation."[3]

Be this as it may, between the pull from world events, and the internal pressure of engineers' philosophical concerns, there are prospects for the profession to rise to new heights. If there is a "power elite" in our society (to use the term coined by C. Wright Mills in 1956) then engineers should be represented in it. But how do we move from rhetoric to reality?

## ENGINEERING AND THE CONCEPT OF THE ELITE

Look up the word in any dictionary.

*Elite:* "The choice part; a superior group."

*Elite:* "The choice or best of anything considered collectively."

*Elite:* "The choicest part, particularly of a society; the pick; the flower."

It's not a word with which we're especially comfortable, either as Americans or as engineers. We tend to make fun of it. (An old radio comedy program opened with Archie the bartender answering the telephone: "Duffy's Tavern, where the elite meet to eat!") Yet, it's an interesting word, *elite,* and I believe that by thinking about it we can learn some important things about ourselves—both as Americans and as engineers.

Appropriately, the word comes to us from the French; it stems from the verb *elire* which means "to elect," "to choose." I say appropriately not only because the French have a reputation for thinking of themselves as the elite among nations, but mainly because they were the people who first conceived of engineering as an elite profession. As far back as 1675, during the reign of the Sun King, Louis XIV, the French Army created a special organization of military engineers—the *Corps des Ingénieurs du Génie Militaire.* At that time the term *ingenieur,* which previously had signified a craftsman-builder, began to take on professional connotation.

In 1716, early in the reign of Louis XV, the French government established a civilian engineering corps to oversee the design and construction of bridges and roads. This organization was called the *Corps des Ponts et Chaussées.* A school to train this corps—the *École des Ponts et Chaussées*—was founded in 1747. This was not a vocational school. It was an engineering school. The French perceived that the findings of science could be applied to practical enterprise, and they acted on this notion. They introduced the study of mathematics and physics, not only in connection with bridges and roads, but also canals, water supply, mines, fortifications, and manufactured goods. The French deserve credit for deciding that these scientific applications should be taught in a school setting rather than simply picked up "on the job." Thus were a number of time-honored crafts transformed into the profession of engineering.

As the eighteenth century gave way to the nineteenth, other technical schools were founded in France, most notably the world-renowned *École Polytechnique.* The importance of state-sponsored technical training was recognized by leaders of all political persuasions. (The *École Polytechnique* was conceived under the monarchy, opened by the revolutionary government in 1794, sponsored by Napoleon, and supported by every government thereafter.) Further, French planners agreed that if the undertaking was to be effective, the students should be citizens of the highest quality. Thus the famous mathematician, Laplace, wrote that the *Polytechnique* should aim to produce young people "destined to form the elite of the nation and to occupy high posts in the State."[4]

The roots of French engineering lie in government support and government service. As private enterprise developed, however, the profession expanded its horizons. The *École Centrale* was founded in 1829 to provide industry with engineers comparable to those who were being turned out by the *Polytechnique.* In the private sphere as well as in the public, the profession maintained a high level of pride and communal esteem. A prominent American engineer traveling to France in the late 1920s was fascinated: "An outsider is impressed," he wrote, "by the prestige of the engineering profession, its important role in the high administrative ranks of State and industry, the personal distinction of its members and the attraction it holds for the best endowed youth of France."[5]

The current state of affairs is much the same. Lawrence P. Grayson, past president of the American Society for Engineering Education, has written that "In France, most of the leaders of business and government have graduated from the elite *Grands Écoles;* these approximately one hundred schools concentrate primarily on teaching engineering and technology."[6] Historian Cecil Smith has paid tribute to the "prowess" and "collective influence" of French engineers in leading their nation to recovery after World War II. He credits them with acting "as planners, economists, urbanists—'inter-ministerial generalists,' drafting legislation and then the decrees to implement it." And further: "Beyond the state administration, the same influence spread throughout the greatly enlarged para-public sector of electric power, gas, coal, banks, airlines, telecommunications, Renault, the atomic energy commission, and into their quasi-private affiliates." And further still: The polytechnic engineers "goaded outmoded firms to merge and modernize for international competition" and then "went out to manage them by the light of systems engineering and operational research."[7]

I do not propose the French model as applicable to the United States. For one thing, the centralized technocratic approach of the elite French engineers has produced notable failures as well as successes. The so-called "polytechnicians" were committed to building canals long after they should have been building railroads. Then their centrally planned railroad system, radiating from Paris, proved less practical than those systems that developed helter-skelter in other nations where free enterprise was dominant. Even today, French railroads have been electrified and modernized at great expense only to end up underutilized.

The polytechnicians have not only made many of the arrogant mistakes that are inherent in central planning, they have played a role in French politics that has come to represent an insular "establishment" approach. During the widespread student riots of 1968, a slogan was painted on many walls that read, "Death to the technocrats." Afterward, the teaching of engineering was introduced at a number of universities where it had never been taught before, and now as many as 20 percent of French engineering students are trained outside of the polytechnic system. Yet, the technocrats were not put to death. In France, the polytechnicians still reign supreme.

In Britain (as I noted in discussing Prince Charles's speech at Harvard, Chapter 6) the evolution of engineering was a very different story. The English upper classes believed in classical education. Elder sons inherited titles and oversaw their family estates. Younger sons sought careers in the church or the army. There was no royally sponsored corps of engineers. There was no meaningful government funding of higher technical education until Parliament, in 1889—more than a century later than the French—authorized grants to city universities for the purpose. Cambridge and Oxford followed reluctantly, with Cambridge introducing a program in "mechanical science" in 1890 and Oxford establishing a chair of "engineering science" in 1909. As I have said earlier, one could argue that Britain's decline as a world power is attributable to the failure to appreciate the importance of engineering education.

And yet the industrial revolution started in England. While French technology was tied to government support and bureaucracy, the British pattern was one of individual ingenuity and entrepreneurial initiative. Private family establishments controlled the making of iron. Free enterprise gave birth to the steam engine and many other mechanical marvels. Knowledge was gained pragmatically, in the workshop and on construction sites, and engineers learned their craft—and such science as seemed useful—by apprenticeship. British engineers did not ignore mathematics and science, nor did the French totally disregard the lessons of experience. But the differences in the two nations' approaches to engineering is striking. Each country—and each engineering culture—has had its successes and its failures.

Yet concerning the social position of the engineer, there is simply no contest. In France, engineering became associated with professional pride and public esteem, and with leadership at the highest level. In Britain, engineering was considered a "navvy" occupation. (The original navvies were laborers on canal construction jobs.) There were, to be sure, notable British engineers, some of whom became famous even to the point of being buried in Westminster Abbey. They were heroic—celebrated, for example, in the verse of Rudyard Kipling—as were soldiers, explorers, and other builders of empire. But in the overall social structure, they were nowhere near the top.

We Americans are heirs to both traditions. As a British colony we inherited a large dose of British class prejudice. The snobberies of Oxford and Cambridge were transplanted to Harvard and Yale—where even today you will have a hard time finding an engineer, or anyone to say a kind word about engineering. But the follies of the British aristocracy are only a small part of the picture. In the early days of the United States, there were so few engineers—fewer than thirty in the entire nation when the Erie Canal was begun in 1817—that Americans had no choice but to adopt the British apprenticeship model. The canals and shops—and later the railroads and factories—were the "schools" where surveyors and mechanics were developed into engineers. As late as the time of World War I, half of America's engineers were receiving their training "on the job."

But the polytechnic heritage also came into play. The United States Military Academy—West Point—was founded in 1802 as a school for engineer officers. Sylvanus Thayer, appointed as superintendent in 1817, visited the *École Polytechnique* and other European schools, and introduced to West Point a curriculum based on what he found there. B. Franklin Greene did likewise in the 1840s when he took over as director of the newly established Rensselaer Polytechnic Institute. These two schools turned out most of the academically trained American engineers of the first half of the nineteenth century, a time when engineering attracted "many sons of leading American families," sons of physicians, attorneys, judges, clergymen, academics, and wealthy businessmen.[8]

When General Sylvanus Thayer, in his old age, decided to endow an engineering school at Dartmouth College, he conceived of it as a two-year graduate program for students who had already completed a four-year college course. He believed that engineers should be "gentlemen" before they embarked on professional education. He desired that the Thayer School, as it was named, should seek exceptional individuals and train them "to become ornaments to the profession." He expected the school to prepare engineers "for the most responsible positions and the most difficult service."[9] This sounds much like Laplace waxing rhapsodic about the first students at the *École Polytechnique*.

But while Sylvanus Thayer, in 1867, was envisioning gentlemen-engineers who would take leadership roles in American society, the dynamics of American life were changing. Railroads were spanning the continent, and industrial development was booming. Technically trained people were needed in large numbers, and nobody cared if they were gentlemen or if they expected to become ornaments to the engineering profession. Quite to the contrary: The Civil War had just ended, and the prevailing values were those of the frontiersman, the immigrant, and the hustling entrepreneur.

Perhaps the most crucial event in the social history of American engineering was the passage by Congress of the Morrill Act—

the so-called "land grants" act—in 1862. This law authorized federal aid to the states for establishing colleges of agriculture and the so-called "mechanic arts." The founding legislation mentioned "education of the industrial classes in their several pursuits and professions in life."

With engineering linked to the "mechanic arts," and with engineers expected to come from "the industrial classes," the die was cast. American engineers would not be elite polytechnicians. They would not be gentlemen attending professional school after graduating from college, as proposed by Sylvanus Thayer. Engineering was to be studied in a four-year undergraduate curriculum. And, as anyone might have predicted, when engineering became increasingly complex and demanding, liberal arts studies within this curriculum would be reduced to the verge of insignificance.

As industrial growth accelerated, the need for engineers grew apace, and engineering schools multiplied and expanded. Thousands of the new graduates went to work for large corporations, thus adding to the problems of maintaining professional status. A few of the engineers in industry rose to positions of leadership, and these joined with the independent consulting engineers in developing numerous professional societies. They fought valiantly to establish a high level of professional culture, and indeed a backward look at American engineering leaders reveals much nobility and distinction. But they were swimming against the tide.

The undergraduate nature of engineering education set engineering apart from and somewhat below such occupations as law, medicine, pure science, and eventually many others, in which graduate education was the norm. Engineering students did indeed come mostly from farm and blue collar families, as had been predicted by the authors of the land grants legislation. (This was true up to almost a generation ago, and even now the tradition—an honorable tradition, I hasten to say—has not totally disappeared.) Most important, the enormous growth of American industry, and its enormous appetite for engineer-employees, many of whom were given sub-professional jobs, all but overwhelmed the efforts of engineers to establish and enhance their position in society.

When, early in this century, the legal and medical professions reconstituted themselves—their schools, their licensing, and so forth—some engineers sought to do likewise. But the prevailing attitude was summed up by W. E. Wickenden in his mammoth *Report of the Investigation of Engineering Education, 1923–1929:* "There would be no conceivable gain to society," he wrote, "in making scientific technology the monopoly of a restricted professional group, as in medicine and law, nor is there any inherent basis for limitations of numbers in technological education."[10]

So American engineering developed, as did American engineering education—as, indeed, did the nation—in a random, vigorous, serendipitous way. Elitism was not a part of the profession's ethos. Indeed, there were occasional debates in the professional literature about whether engineers ought to join labor unions. Several of the engineering societies were battlegrounds between employee-engineers, executive-engineers, government engineers, and independent consulting engineers. Efforts to bring all of these together into a single professional organization repeatedly ended in acrimony and fragmentation. One might not think so intuitively, but engineers happen to be an extremely contentious group! That has only added to the difficulties of trying to achieve esteem and status as a profession.

When C. Wright Mills wrote his widely read book, *The Power Elite,* in 1956, he reported that engineers were typically reduced to the role of "a hired technician," with true power being vested in "the corporation chieftains and the political directorate."[11] That was more tactful than Thorstein Veblen had been in 1917 when he wrote that the public viewed engineers as "a somewhat fantastic brotherhood of over-specialized cranks, not to be trusted out of sight except under the restraining hand of safe and sane businessmen." "Nor," he added, "are the technicians themselves in the habit of taking a greatly different view of their own case."[12]

The situation today must distress anyone who cares about the profession of engineering in the United States. As I have observed earlier, in the political arena engineers are almost invisible. In government bureaucracy there are many engineers, but toward the top of the power pyramid their proportionate representation declines.

In the media, and in the arts, engineers have no voice. In industry, leadership is vested in MBAs, accountants, and lawyers, except in companies where engineering is the heart of the process, and even there it is often the moneymen who rule. Engineers generally earn less than the other professionals with whom we like to compare ourselves. Judged by practically any measure, engineering is not where the power is.

According to the old adage, engineers are always on tap, rarely on top. A 1985 book about the engineering profession was entitled, *Mechanics of the Middle Class*.

Yet, the work of engineers is central to much that is of supreme importance in our society. The work of engineers is a key element in our hopes for a better life, for ourselves and for all of humanity. In this sense, engineers have "behind the scenes" influence that provides certain psychic rewards. Also, engineers find in their creative work much genuine satisfaction—what I have called existential pleasures—making them in some ways the most fortunate people in the world. American engineers have served their nation well and have been compensated for their efforts, adequately if not lavishly. Some might say that the situation is not so bad, either for American engineers or for American society. Some people *do* say that we ought to be cautious about tinkering with a system that has worked so well.

But many observers—an ever-increasing number—believe that the situation is not at all satisfactory and that something should be done to change it. The argument is not subtle or difficult to grasp. Let me repeat it in its simplest terms: We live in a technological age, and if our society is to flourish, many of our leaders should be engineers, and many of our engineers should be leaders.

The argument is given force not only by inner logic but also by what we see around us. In West Germany, a majority of the corporate leaders are alumni of the technical universities. The leading role of the engineer in France has already been discussed. In Japan's leading companies, more than 65 percent of the members of boards of directors have graduated from science and engineering programs. In Taiwan, most leadership positions are in the hands of scientists and engineers, many of whom were educated in the United States.

As of the time of this writing, the presidents of Chile, Peru, and the Phillipines are engineers. The prime minister of Italy—disheartened by the failures of conventional politicians—has named a cabinet that has among its twenty members four scientists and three engineers.

The president of China, Jiang Zeman, is an electrical engineer who rose to political power through the post of minister of the electronics industry. President Jiang recently told an American visitor that although he no longer uses his engineering actively, it helps him in his thinking and understanding, not only of commercial and industrial projects in his country, but also in systemizing his thinking about the world.[13]

These facts have become a cliché of economic and political discourse. So, too, has America's declining economic competitiveness and the failure of American technology to stem this decline.

How are we to respond to the urgent need for enlightened, savvy engineering leadership? How are we to address—in today's changing social and economic climate—the question of engineering and elitism?

One thing that we do *not* want to do is to get hung up on the question of titles and honorifics. I've heard the tales of American engineers who travel to Latin America and are pleased to find themselves addressed as "Ingeniero." I vacationed in Portugal a few years ago and noted in a travel guidebook that engineers there should properly be addressed as "Senhor Engenheiro." I thought that was splendid until I read further and found that "Indeed, every university graduate is called Senhor Doutor, so it does no harm and is flattering to do so if in doubt." The history of German engineering is full of battles about who was entitled to be called what depending in large part upon whether the individual in question had studied Latin. It reached the point where one engineer said about some of his colleagues that "the kind of narrow-minded status conceit and petty addiction to titles achieved here cannot be trumped even in the land of titles and ribbons."[14] With all due respect and affection for our Latin and Continental colleagues, Americans do not take too kindly to titles and ribbons. If engineers press too far along these lines we are liable to end up embarrassing ourselves.

On the other hand, a wholesome respect for honors can be a force for great good. We have to draw a fine line between petty vanity and legitimate pride. The ancient Greeks placed a high value on personal reputation, and indeed the great Greek tragedies were entered into competitions for which prizes were awarded. It is true that Jean-Paul Sartre declined to accept the Nobel Prize for Literature in 1964 on the ground that honors tend to subvert commitment, but his view—although we can appreciate his point—is not widely shared. Titles, honors, and prizes, used with discretion, can reward excellence, show respect for excellence, and hence inspire people to strive for excellence. One element in the rise of engineers to a higher level of regard in our society might be an increased regard for individual reputation—reputation based on accomplishment and intended to encourage greater accomplishment.

I think it totally appropriate that the Draper Prize was established in 1989 to serve as an equivalent to a Nobel Prize for engineering. The first was awarded to Jack S. Kilby and Robert Noyce for their invention of the integrated circuit. In 1991 the award went to Sir Frank Whittle and Hans von Ohain for their development of the jet engine. In 1993 the honoree was John Backus, the inventor of Fortran, the computer programming language used in most engineering work. The 1995 award was presented to John R. Pierce and Harold A. Rosen for pioneering development of communication satellites.

It is splendid and inspiring when the National Medal of Technology is awarded to an outstanding engineer or when a creative genius is inducted into the National Inventors' Hall of Fame. I support the aims and activities of the engineering honor society, Tau Beta Pi, and I get a thrill from reading about the multitalented young men and women who are selected each year as its laureates. It is reassuring that the professional engineering societies are dynamic, and that issues of accreditation, curricula, degrees, and licensing are earnestly debated, along with questions of ethics and professional responsibility. It's truly significant that there is a National Academy of Engineering. (It is also significant that it was not established until 1964, a full century after the founding of the National Academy of Sciences.) I think it particularly admirable that the members of this academy have started to speak out on matters of societal concern. (More on this in Chapter 12.)

These developments are encouraging. They herald changes in the image of the profession—not merely a public relations transformation, although there is admittedly some PR involved (and why shouldn't there be?)—but changes in image that are based on a meaningful evolution in the profession's talents and concerns. When people think about the leaders of American society—the elite of American society—they may start to think about engineers.

However, engineers pride themselves on being realistic. This being so, they must recognize that if engineers are to become leaders and leaders to become engineers, far-reaching changes will be required, changes that go beyond those just mentioned. Most particularly, engineering education must be revitalized, liberalized, and enriched. If we want to develop renaissance engineers, multitalented men and women who will participate in the highest councils, we cannot educate them in vocational schools—even scientifically distinguished vocational schools—which is what many of our engineering colleges are in danger of becoming. I will return to this topic in the next chapter.

Also, while we're being realistic, let's think a little bit about the people who are engineers in this country and those youngsters who want to become engineers. They are not all fervently ambitious polytechnicians, nor do all of them want to be. Nor, to be honest, is that what we expect or even want for all of them.

As I noted in Chapter Seven, a recent sociological study suggests that a third of American engineers constitute a "rank and file," seeking only "a modest level of technical challenge combined with the opportunity for periodic promotion and their share of organizational recognition." We must admit that this statement has the ring of truth. And who among us dares reproach these people for their modest ambitions, much less suggest that they're not entitled to be members of a "greater" engineering profession? The study also shows that many engineers seek career satisfaction mainly in technical work, while others seek managerial success in terms of a traditional climb up the corporate ladder. In fact, there is little in the study to indicate that the average working engineer even dreams of taking on a leadership role in society. Other studies and surveys reveal similar patterns. Apparently the idea of engineering prominence in society, although championed by leaders of the profession, has not enticed the average practitioner.

So, after this harsh dose of reality, what are we to do? I suggest that we should be pursuing two courses of action. First, we must build on all the good things we know about our profession, trying as best we can to raise the general level of professionalism for all engineers. A rising tide to raise all boats.

Second, we should seek out and encourage those exceptional engineers—of all ages, specialties, and geographical locations—who have the talents and aspirations to take on "the most responsible positions and the most difficult service," to quote Sylvanus Thayer. Jerrier Haddad, the retired IBM vice president who chaired the 1985 National Research Council report called *Engineering Education and Practice in the United States,* suggested that perhaps 10 to 15 percent of the nation's engineering students—those who show special promise—might be offered an enriched and lengthened curriculum.[15] Even within the standard four-year program, there are ways to encourage cultural and intellectual breadth along with other attributes of leadership. And while we are encouraging leaders *within* the profession, we should be trying to attract leaders *to* the profession. We should be trying to make engineering more appealing to potential entrepreneurs and political activists. Also to intellectuals, and particularly to idealists.

Is that too daring an idea? Does paying special attention to a small minority of the profession carry elitism too far? An elitism based on talent and ambition is very much in tune with the American temper. We believe in equality of opportunity, freedom of choice for the individual, and a firm disavowal of snobbery. Given these preconditions, however, we admire outstanding achievement. We thirst for leaders, for people with extraordinary talents and ambitions. It is not inconsistent to strive to become the greatest profession we can be—including a "rank and file" worthy of esteem—and at the same time to encourage a few of our number to rise to special heights. After all, not every lawyer rises to a position of eminence, and only nine are justices of the Supreme Court.

I see the cultivation of engineering leaders as a source of pride and satisfaction for all engineers, not in any way a cause of discord or resentment. It's a wholesome aspect of human nature, I believe, to applaud the successes of one's co-professionals. We engineers are a fellowship—a brotherhood, and increasingly a sisterhood. We root for each other much as we root for our nation's athletes in the Olympic Games. We also root for each other because we know how much good successful engineers can do for society.

Having said all this, I still feel an element of discomfort, and I believe it relates to the word itself: *elite.* It doesn't have the right ring to it. It doesn't resonate to the heritage of American engineering. Maybe we'll never be comfortable with that fancy word with its French origins. But even if we are uneasy with the word, even if we eventually discard it, we should hold tight to the concept. We need people who are "the elect," "the pick," "the flower" of our society. We need people who feel "chosen," chosen for leadership and service.

In a spirit of wholesome pride and love for our profession, and keeping in mind our nation's need for enlightened technological leadership, let me once more quote the remark that Laplace made upon the founding of the *École Polytechnique.* You will recall that he looked forward to educating young people "destined to form the elite of the nation and to occupy high posts in the State." It's a somewhat rhetorical flourish, from a man who lived in another land in another age. But the statement can brighten our day and help us to keep our sights high—which is something that engineers like to do.

CHAPTER 11

# EDUCATION

## AGENT FOR CHANGE

Let me repeat: It is important that engineers—or at least a goodly number of engineers—come forward to play a more prominent role in American society. In order to do this to best effect, they should be not only smart (which they already are) and honorable (which most of them are), but also broadly educated, astute, and wise (which is a tall order yet a worthy goal). It is also important that engineers become more effective in a technical sense as well: creative and competent, able to keep the United States in the forefront of the industrial world. The nation's well-being, humanity's well-being—and perhaps at this point even the planet's well-being—depend upon the vitality and advancement of the engineering profession.

Having now said this a number of times in a number of ways, the question then becomes: What are we going to do about it? Like many advocates, I tend to rely on rhetoric. "It is important that" we do this or that. We "must," we "ought." Exhortation becomes a way of life.

Not that I underestimate the power of words. Martin Luther posted his ninety-five theses on the door of the castle church at Wittenberg, and the world changed. But, in Luther's time (1517), the world was ready to change and there were a lot of people ready to change it.

In what ways are we ready to change today?

When Jeanne Kirkpatrick, historian and diplomat, was asked why so many foreign affairs experts were taken by surprise by the collapse of Communism in Eastern Europe, she said they had failed to consider what can happen when a new generation comes to maturity and addresses the appalling problems that confront it. She also acknowledged underestimating the practical effect of action by a single resolute individual such as Mikhail Gorbachev.[1]

I heard a similar question asked of Richard Helms, one-time director of the CIA. How was it possible that our gigantic intelligence agency had failed to predict such a momentous event as the falling of the Berlin Wall? "Well," replied Mr. Helms, "you can only expect spies to tell you what people are planning to do *when the people themselves know what they are planning to do."* The people who tore down the Berlin wall didn't know that they were going to tear down the wall until just before they did it.[2]

Thus, three agents for change: (1) A new generation sizing up the problems that confront it; (2) The endeavors of extraordinary individuals; and (3) The spontaneous actions of large groups, triggered by incidents almost impossible to predict. To this list we should certainly add: (4) The reasoned strategies of thoughtful professionals, for example, the containment policy proposed by State Department specialists and carried out by the U.S. government against Communist nations for almost half a century.

What will bring about change in the world of engineering? How will we produce the engineers so sorely needed?

Sometimes I feel that the revolution is already under way. I read an article about the fervor of young "hackers," and fancy that the computer will be the agent of change, not only in a broad societal sense, but in creating enthusiasm for engineering as a profession. I hear a speech at an engineering symposium, read an editorial in a professional journal, or even an impassioned letter to the editor, and I think, yes, this is the clarion call we have been waiting for. I note that the Clinton administration comes in on a platform calling for "change," Vice President Gore spells out the importance of technology as a national priority, and the proposed budget begins to reflect, however slightly, this cast of mind. Republicans cut the budget, but at the same time swear allegiance to technological progress. House Speaker Newt Gingrich stresses the importance of the information revolution, and rhapsodizes about the prospects for space travel. Maybe we are already in the midst of societal transformations that we don't even recognize, occupied as we are with political gossip and other diversions.

Then there are the small events that one just happens to notice, seemingly trivial but perhaps making all the difference. Engineering faculty at Iowa State University, finding that 92 percent of the state's first graders think engineers drive trucks or trains, or work on cars and engines, decide to take action. Working with an elementary school teacher, they develop a curriculum supplement called *Not All Engineers Drive Trains*. Consisting of a teacher's resource manual, an illustrated storybook, and a coloring book (and financed by Rockwell International and the National Science Foundation), the package has been enormously successful. We hope for tall oaks from little acorns.

This last example reminds me that surely the greatest of all agents for change is education. By creating the engineers of the future, educators can transform the world in the most meaningful way possible. Yet engineering education cannot flourish in the absence of popular regard and government support. We have something like a Catch-22 here. Appropriate education is needed to further a renaissance, but a renaissance in engineering is needed to inspire steps toward appropriate education. Someone must break this paralyzing cycle. Engineering educators, as a group with actual (albeit limited) powers, are a likely choice for taking the initiative. Unfortunately, engineering educators, inundated with complaints and undersupplied with resources, tend to be a conservative group, more than a little skeptical about possibilities for reformation.

Yet, as the French philosopher Henri Bergson once observed, it is easy to prove the impossibility of acquiring any new habit: "If we had never seen a man swim, we might say that swimming is an impossible thing, inasmuch as, to learn to swim, we must begin by holding ourselves up in the water and, consequently, already know how to swim." "But if," proposed Bergson, "quite simply, I throw myself into the water without fear, I may keep myself up well enough at first by merely struggling, and gradually adapt myself to the new environment: I shall thus have learnt to swim."[3]

We lift our spirits by resorting to philosophical metaphor. We raise our hopes by alluding to historical analogies. But these cannot substitute for action. We engineers cannot in good conscience exhort the public, admonish the government, and in general toot our own horn, unless we devote ourselves to the necessary labors. School is where it all begins, so school is one place where we must concentrate our efforts.

The earlier in school one can reach young people the better. The Iowa engineers who target first graders have a wonderful idea. The concept of "technological literacy" for all—an awareness and appreciation of engineering, its technical appeal, and its social importance—is the best possible underpinning for the changes we seek. But there are limits to what engineering professionals can do to change K through 12 education. Whatever they *can* do they should certainly *try* to do. A splendidly constructive activity takes place each year during Engineers Week (initiated in 1951 by the National Society of Professional Engineers, and now sponsored by all the major societies) when thousands of engineers go into their local schools as missionaries on behalf of their profession.

It is interesting to note that Engineers Week is scheduled each year in February to honor George Washington's birthday. The connection is that Washington, in addition to his more justly famous accomplishments, spent several years as a land surveyor. I've always thought it was a bit of a stretch to imply that this makes him a quasi-engineer by profession. On the other hand, it is good to remember a time when so many of our nation's greatest figures were, by inclination and training, technologically erudite. (Ben Franklin and Tom Jefferson—where are you in our moment of need?)

Outreach programs are extremely important, particularly in the elementary and secondary schools. But the place where the engineering profession has the most effective control, and can make the most prompt and productive changes, is in the engineering schools. These institutions are both the source of many of our problems and the key to potential solutions.

OUR SCHOOLS TODAY

We are engineers, not idle dreamers, so we must start with what we have: approximately 330 universities, colleges, and institutes with engineering programs accredited by ABET (the Accreditation Board for Engineering and Technology, which is governed by the major engineering societies). Each of these institutions has its own history and traditions, its own areas of emphasis and specialization. Some are huge university establishments with celebrated research facilities and graduate programs. Others are small undergraduate colleges. All share certain basics required in the ABET-accredited four-year undergraduate curriculum: specified amounts of mathematics and the fundamental sciences, plus an assortment of basic engineering sciences, and a minimum 12-1/2 percent component of liberal arts courses (*i.e.*, the equivalent of one full semester of courses in an eight-semester degree program).

Most discussions of engineering education fail to take into account the astonishing diversity among the various schools, based not only on the factors just mentioned, but even more on differences relating to geography. Each institution has a synergistic relationship with its local community, drawing many of its students from the area and sending many of its graduates to work in local industries. The affiliations between engineering schools and industrial neighbors are often so intimate that they partly define the nature of the educational process. For example:

> At Purdue most students find jobs within a three hundred-mile radius of the Lafayette, Indiana, campus. The biggest recruiter is General Motors, and many students come from automotive families. A large Delco Division plant making integrated circuit chips for automobiles is only thirty miles from Lafayette. There are jobs in heavy manufacturing south of Chicago. Students interested in electronics favor such firms as Magnavox, Zenith, Northrop, and Motorola, which have plants near Chicago in an area that Purdue students call "Corn Valley."

UCLA students find themselves in an area heavily populated by defense contractors—Hughes, TRW, McDonnell Douglas, Lockheed, Rockwell, and Northrop, to name a few. Forty-five percent of UCLA engineering graduates stay in southern California, and for them the defense sector, and in particular the aerospace industry, has significant appeal.

Stanford students are attracted to such Silicon Valley companies as Intel, Advanced Micro Devices, and Apple. The most popular large firms are Hewlett-Packard and IBM.

The University of California at Berkeley is about midway between Silicon Valley and the Lawrence Livermore Laboratory, and there is considerable discussion on campus about work in the private sector versus work in government-sponsored research.

At Iowa State the leading local employers are John Deere and Caterpillar Tractor.

At Oregon State, 60 to 70 percent of the graduates find work near home, either in the forest products industry or with such local high-tech firms as Intel, Hewlett-Packard, Floating Point Systems, Tektronix, and Mentor Graphics.

Most University of Illinois graduates join large companies within the state. Some are attracted to California, and others to a major McDonnell Douglas facility in St. Louis.

More than 60 percent of Georgia Tech's students come from within the state, and most of these find work there. (At other southern schools the percentage of locals is typically much higher.) Georgia Tech has a major interest in manufacturing systems, and many graduates follow that specialty, favoring such long-established companies as Dow, Proctor & Gamble, DuPont, IBM, and GE.

At the University of Maryland a very large proportion of the students come from the District of Columbia area and most want to stay there. 60 to 70 percent take jobs either with local high-tech firms or with federal agencies.

Three quarters of the students at RPI, in Troy, New York, come from the northeast and find work in the northeast. Major employers are IBM, GE, Raytheon, Digital

> Equipment, General Motors, United Technologies, and AT&T.

Many other examples can be cited, such as the connections between southwestern schools and the oil industry. There are also exceptions to prove the rule, such as MIT which draws its students from all over the world. But even at MIT, about a quarter of the graduates stay in Massachusetts.[4]

Diversity is a source of strength; but it also presents problems when important profession-wide issues are up for discussion. A major research university, with close ties to eminent high-tech corporations, has a world view quite different from that of a school with modest reputation and troubled finances, accustomed to placing graduates in low-level jobs with local manufacturers. Nevertheless, the essentials of a basic engineering education must be the same nationwide, or else the term "engineering profession" loses all meaning.

## THE NEED FOR IMPROVEMENT

American engineering schools are esteemed worldwide, as evidenced by the many ambitious young people who come from abroad to study in them. But the shortcomings of American engineers, as touched on in previous chapters, and as widely discussed by engineers and employers of engineers, require attention. This means that engineering education is in need of constructive change. Just about everybody agrees on this, even the many cynics who question whether deeds will follow rhetoric. One used to hear engineering educators saying "If it ain't broke don't fix it." But I haven't heard that expression used in this context for quite a while.

From 1991 to 1995 I was privileged to serve on the Board on Engineering Education (BEEd) established by the National Research Council. We were charged with "identifying significant issues in engineering education; facilitating communication about engineering education needs among academic, industrial, and government leaders; developing long-term strategies . . . formulating timely policy recommendations; and stimulating actions to implement the strategies and policy recommendations." In pursuing these objectives, the board held about a dozen meetings with numerous representatives of industry, academe, government, and the professional societies, and followed these meetings with four symposia in various parts of the country. The result was a report entitled, *Engineering Education: Designing an Adaptive System,* which I commend to anyone interested in the topic. Essentially, the report set forth needs and goals as enunciated by participants in the process and evaluated by members of the board, and then—the key element—recommended that each engineering institution undertake self-assessment in the light of these needs and goals. Self-assessment is to lead to effective change.

Although I participated with other board members in drafting the report, my most vivid memories are of the meetings and symposia that preceded it. The oral presentations—often passionate beyond what one might expect, given the topic—carried a punch that was impossible to capture on paper, particularly in a government-commissioned document.

## HIT THE GROUND RUNNING

The most fervent appeals came from representatives of industry who wanted engineering graduates trained in practical ways. They complained that current curricula are too heavily weighted with theory, making for engineers who are poorly equipped to enter industry and produce. A particular grievance related to "manufacturing," which business executives felt was slighted in engineering school with disastrous implications for American industry. They also decried the neglect of "design" and the stress on "theory," the neglect of "synthesis" and the stress on "analysis" or simple problem solving.

Indeed, theoretical studies, rather than hands-on practice, have for half a century been the heart of engineering education. There are historical reasons for this as well as academic. After World War

II, engineering leaders recognized that scientists had increasingly taken over many aspects of engineering work because engineers had inadequate mathematical and scientific training. So a deliberate attempt was made to rectify the situation, spurred on by the influential "Grinter Report" of 1955.

Modern engineering is based in large measure on mathematics and the sciences, and nobody seriously proposes that these topics be dropped in favor of a trade school approach. Yet, there is ample evidence that the pendulum has swung too far in the direction of pure theory, and the consensus is that correction to a more appropriate balance is needed. The gap between engineer and technical worker has been partly filled by graduates of four-year curricula in "Engineering Technology," where the hands-on approach is stressed. There are also "Technician" courses, mostly two years in length, centered in a large network of community colleges. Nevertheless, the demand for engineers with practical training continues to grow.

Any attempt to change is complicated by the engineering research enterprise that has also evolved over the past half century. University research, funded mostly by the federal government, has yielded marvelous results. But one unfortunate side effect has been the growing financial reliance of engineering schools on funded research, and the evolution of faculty incentives along these lines. As the pressure to obtain research grants intensified, the attention paid to teaching inevitably diminished. Graduate students, in order to gain financial support, become of necessity research assistants. Thus there was an insidious trend away from the sort of teaching that most people favor, teaching that features a healthy mix of theory and practice. To correct the imbalance, some schools are introducing experimental changes in curriculum; also, faculty incentives are being addressed, notably by National Science Foundation grants for excellence in teaching as well as research. But, lacking additional financial resources, it is easier to identify the problems than it is to cure them.

It is also easy to overreact, and I fear that some industry people are doing just that. Looking for graduates who can "hit the ground running" (a phrase that I found increasingly annoying as the BEEd hearings went along), they tend to underestimate the importance of mathematics and the sciences while overestimating the value of the practical hands-on training that students can get in engineering school.

I'm a construction engineer, probably one of the least theory-related sorts of engineers there is—a hardhat really, a muddy boots kind of person. From the first day on my first job I discovered that there were a lot of practical things I wasn't taught in engineering school, but they were hardly ever the sorts of things that *can* be taught in a professional school, or even ought to be.

I will never forget the time, on one of my first jobs, that I was told to obtain wooden shores to support forms for a concrete pour. I did some quick calculations, checked my handbook for information on stresses and strains, and then ordered out a truckload of six-by-six timbers, ten feet long. I thought that these were perfect for the prescribed use, and I was quite proud of the way in which I had handled my assignment. However, when the material arrived at the construction site, I received a phone call from the project superintendent who was furious. "Who is the damned fool," he yelled, "Who is the damned fool who ordered out these goddam timbers?" I said that I was the responsible party, and asked what was the problem. "Don't you know we use four-by-fours, not six-by-sixes?" he said. "But," I responded, "that would have required a lot more pieces. I thought this was the best design solution." "You idiot!" he yelled. "The six-by-sixes weigh a hundred pounds, and a single carpenter refuses to handle them. The four-by-fours weigh less than half as much. That's why we use four-by-fours and never six-by-sixes. You want to come out here and try to put these damned tree trunks into place?"

This was one of the many things about construction that I wasn't taught in engineering school. And just about every day throughout my working career I have learned at least one more lesson of the same sort. But if in college I had been studying the nuts and bolts of how things work on a construction job, I wouldn't have had time to study the science-based subjects that made up most of the curriculum. My professors did give me many practical pointers, for which I am eternally grateful; but most workaday tricks of the construction trade I learned where they are best learned—on the job.

I may be a hardhat-engineer-businessman, but I don't regret the time that I spent studying physics and chemistry, mathematics and electricity, mechanics and thermodynamics. (Well, maybe I could have done with a little less thermodynamics!) I still have a feeling for force vectors and the stresses and strains in structural members, for the chemistry of materials, the flow of electrons, the behavior of fluids, the relationship in gases of pressure and volume and temperature. And I believe that the engineering courses I took those many years ago, mostly forgotten and partly obsolete, still help me think more clearly, and are of value to me in ways that I find hard to put into words. I remember many of the things that my professors taught me—not the details, but rather the concepts—the basics of how to approach problems, how to *solve* problems. I know many engineers in my field, and of my generation, who feel the same way.

I also find in engineers, regardless of age or specialty or level of technical sophistication, a *pride* in having studied engineering, in having worked hard and understood much, even if most of the subject matter was never put to practical use (defining practical in a narrow sense). This pride in a rich, basic education lies at the heart of professionalism, and it shouldn't be undermined in the name of expedience.

Moderation, in the words of Euripides, is "the noblest gift of Heaven." Let us observe it in seeking the right mix of theory and practice in engineering curricula. Let academics recognize that their graduates should be prepared to enter the real world. But industrial employers will not improve the real world by being short-sighted. Investment in some on-the-job training will pay dividends in the long term. And each graduating engineer, we must hope, will be committed to a lifetime of continuing education.

## THE LIBERAL ARTS

Having just quoted one of the great playwrights of ancient Greece, I find myself thinking about the nontechnical part of the engineering curriculum, the so-called liberal arts. This is another area in which the employers of engineers are more than a little ambivalent.

A hardheaded, no-nonsense approach to this perennial problem was expressed by Kent M. Black, CEO of Rockwell International, in a plenary address at the 1993 Centennial Conference of the American Society for Engineering Education. "Not so important," observed Mr. Black, "are some of the social studies, philosophy, English literature, and even history and art." "These are *personal* interests," he continued, "and engineers, who by their very nature are curious, will pursue those subjects that interest them—after graduation." Surprised gasps could be heard throughout the hall, since this is a point of view not usually espoused by today's executives, at least not in public.

A more typical view is that expressed by Hugh Coble, a group president of Fluor Daniel, in a speech to a nationwide gathering of state registration board chairmen. Industry's greatest need, according to Mr. Coble, is engineers with training in the broadening disciplines—humanities, economics, history, and communication—and without such professionals, he believes, "we will fail to regain our leadership role in world engineering and technology."[5] I have written many impassioned pages preaching the same sermon.[6]

Everyone agrees—even the Mr. Black who would eliminate humanities from the curriculum—that today's engineers must be able to communicate. Long-standing deficiencies in this area must be addressed. It is critical that engineers be able to write and speak clearly, listen intelligently, and participate effectively in discussions—with their peers and with people of many different backgrounds and orientation. The old paradigm of the engineer—a specialist working, mostly alone, on a series of isolated projects—is no longer valid. New technologies, such as bioengineering and the information highway, are increasingly multidisciplinary. Long-established technologies, such as the automobile, have changed from relatively simple mechanical devices to complex, electronically controlled, chemically sophisticated systems with ever-changing social and environmental implications. Challenging socio-technical issues, such as arms control, health care, multimedia networks, and infrastructure require systems approaches, interdisciplinary strategies, and group effort. The public is not as naive about technology as it used to be, making the political element of engineering projects ever more consequential. And changes in the world economy require American engineers, more than ever before, to think in terms of global competition—and global cooperation.

There is little dispute about these facts. The problem is that some people, like Mr. Black (and I refer to him only because he expressed in a public forum what others mutter in private)—some people, then, think that communication skills can best be taught in technical report writing courses, while leadership can be inculcated by way of public speaking. I do not agree. The liberal arts are what fill out a person's education, helping turn narrowly focussed professionals into discerning citizens, intelligent communicators, and potential leaders. Courses in technical report writing are not only less effective than literature and history for improving communication skills; they are deadly dull.

This brings us to the heart of what in my estimation is truly the worst defect of engineering education in the United States: The program is laborious and disagreeable.

## ESCAPE FROM ENGINEERING BOOT CAMP

The roots of engineering education in the United States are to be found at the Military Academy at West Point. It was here in 1817 that Sylvanus Thayer, the newly appointed superintendent, introduced a program based largely on what he had learned during a year spent at the *École Polytechnique* in France. I have earlier spoken of Thayer as the champion of broad, liberal education for engineers, and of his efforts, in his later years, to establish engineering as a graduate professional discipline. But, just as he deserves credit for his efforts to bring engineering education to the highest possible level, so must he bear blame for introducing harsh traditions that afflict us today. Or perhaps the blame is ours for not abandoning the harsh traditions when they were no longer suitable for our purposes.

The Thayer System—which has remained the dominant pattern of academic life at the Military Academy—features rigid discipline, intense academic pressure, and almost ruthless "weeding out" of those students who fail to measure up. This approach became pervasive, first as West Point graduates occupied leadership positions in most of the nation's early engineering schools, later as the system became traditional—taking on a life of its own. Discipline and strict moral standards became embedded in engineering education, so much so that the prospectus of a leading engineering school in 1892 contained this admonition:

> Parents who have not governed their sons at home, and inspired them with the principles of sound morality and the spirit of gentlemen, are respectfully requested NOT TO SEND THEM HERE.[7]

This stern spirit affected not only the general atmosphere—arduous assignments, competitive grading, and deliberately high attrition—but also the structure of the curriculum. Instead of trying to interest the students in engineering, to nurture enthusiasm for the profession that was to be their life's calling, the program was designed as an obstacle course. The first two years were dedicated almost exclusively to mathematics and the basic sciences, with no effort made to show how these often brutally difficult studies would help the future engineer do constructive work.

The social ambience has begun to soften somewhat in recent times—although in many schools not nearly as much as one would wish and expect. The Spartan legacy dies hard—"If I had to do it, you can do it!" Also, many junior faculty hail from Asian nations where the cultures stress hard work with no frills. However, even where the general spirit has eased, the curriculum remains forbidding. The idea of giving freshmen and sophomores *engineering* courses—in which the students work together in teams to design and make something—is considered revolutionary.

I am suggesting, however idiosyncratically, that engineering education could be redeemed by making it more *enjoyable.* Most engineers agree, according to studies and polls, that the professional practice of engineering supplies much satisfaction. But if the prerequisite training gives no hint of this, if indeed the study of engineering brings on the opposite—discontent, displeasure, and malaise—then the harm done may be far worse than anything we have heretofore imagined.

The time is past when strict discipline and onerous labor were accepted as the price to be paid for pursuing a rewarding career. Hard work, yes, but that is something different. Today's young people—including some of the very best—will not buy into an education ethos of a bygone age. More's the pity some of us may say; but perhaps we are wrong; and in any event the facts are the facts.

There is an expression that I first heard when I decided to study engineering and that has haunted me all my life: *If you're smart enough to be an engineer you're too smart to be an engineer.* What that means, of course, is if you're smart enough to be an engineer you're smart enough to be just about anything; and if engineering isn't the most rewarding career choice, then why pursue it? In the context of my present argument: If engineering is made to seem unrewarding, indeed distasteful, by the very people who should be revealing its glories, then the smartest students will choose a different outlet for their talents.

Certainly the idea of making engineering the basic course of study for able people—the way it is in France and many other nations—will not take hold in the United States unless we make fundamental changes. If the course of study that once produced intrepid engineers is now perceived as suitable for nerds, then none but nerds will pursue it. I exaggerate—but not excessively—to make a point.

Why is it that women shun engineering even though they have been sought out by the profession for several decades? While they claim their rightful place in law, medicine, business, and many other callings, women have leveled off at approximately 16 percent of the engineering student body. Blacks, Hispanics, and Native Americans are significantly under-represented in engineering schools, and their attrition rate is appalling. Studies show that scholastic ability is not the main problem: Women and underrepresented minorities shun engineering studies because they find them unpleasant. This impairs our society in a number of fairly obvious ways, not least because white males are becoming an ever smaller portion of the working population.

With a recrafted curriculum, introducing real engineering at an early stage and showing it to best advantage, this situation can be improved. Instead of a medicinal dose of math and science, let freshmen get a taste of engineering design (a simple robot, perhaps) and a sense of how engineering fits into society at large. Math and science will then take on added meaning and appeal. With an enriched program, linking technology creatively with the arts and sciences, attrition will decline and there will be fewer reasons for industrial employers—or anybody else—to complain about the quality of engineering graduates. With enthusiastic teaching—supported by an enlightened incentive system for faculty—with proper advisory assistance and a generally nurturing environment, engineering can gain the popularity it deserves, and that our society needs for it to have. Graduating engineers will be better engineers, in large measure because our best young people will be attracted to the profession and will not lose heart while pursuing it.

I speak mainly in generalities, but the world of engineering education is astir with specific activities. The National Science Foundation spends $25 million per year supporting experimental curriculum changes in about sixty engineering schools, grouped in cooperative coalitions of five to ten institutions each. The NSF also supports young engineering faculty who aspire to integrate research with superior teaching, and NSF is funding links between industry and academe, supporting engineers who move constructively between the two worlds. The cover story of the September 1995 issue of *ASEE PRISM* reports on several innovative "introduction-to-engineering" courses designed to attract and inspire freshmen. The same month's issue of *IEEE Spectrum* features an article titled "Educating the Renaissance Engineer."

The NRC Board on Engineering Education report, mentioned above, is chock full of recommendations for engineering schools (self-assessment, modifying faculty incentive systems, improving teaching methods and practices, pursuing curricular reform, encouraging lifelong learning, etc.), as well as counsel for industry, government, and the professional societies. The societies themselves maintain active educational programs, and ABET, the accrediting authority, long considered an obstacle to change, has evinced a new willingness to endorse flexibility.

Another report, *Engineering for a Changing World,* a joint project by the Engineering Deans Council and the Corporate Roundtable of the American Society for Engineering Education, proclaims that engineering education must be "Relevant, Attractive, and Connected." By "Connected" the authors of the report mean to stress integration with the broader community, both inside and outside academia: For example, they propose that each engineering school, in cooperation with local industry, should "partner" with at least one local school at the K–12 level. They also call for across-the-campus outreach, for connections with schools of business, medicine, arts, sciences, and education. But it is here that the prospective reformer stumbles upon one of the most irksome barriers to achieving the worthy goals now so widely shared.

## THE POINTILLIST APPROACH

To nonacademics, nothing seems simpler than the notion of creating a new curriculum. We merely say to the deans and faculty, dear ladies and gentlemen, please proceed. All you teachers of engineering, science, history, language, literature, economics—just get together and do what needs to be done.

It is difficult for the outsider to appreciate how naive this view is, and how complicated the problem really is. Recently I met with an engineering administrator at a fine university—a liberal, progressive institution where one would expect the best sorts of things to be happening. To my surprise, he complained bitterly about the mathematics department. The math professors refused to say even a few words to engineering students about the *applicability* of mathematics to engineering, refused to include in their classes a few examples of how engineers actually *use* math. He wanted them to help his students see that mathematics was not a mere obstacle to be overcome, but a vital element of the profession to which they were committed. But apparently the mathematics professors were determined to teach their subject "straight," pure, on its own terms, as if each student were destined to become a Ph.D. in that field. It was a matter of pride in their specialty and honor for their department. The engineering administrator was utterly disheartened. And if at a liberal institution the math department resists cooperation with the engineering school, I can only imagine how frustrating things are in other places and in other circumstances. It is difficult enough for engineering deans to achieve change within the realm that they nominally control. When they go outside their own domain—as they must do to revamp the curriculum in a meaningful way—the problems seem overwhelming.

*Department.* One little word that represents the source of so much frustration. Departmental structure is an essential element of American higher education; the tradition cannot be all bad, as evidenced by the splendid accomplishments of our colleges and universities. Yet, as we seek to reconstruct the engineering curriculum, the legacy is a disaster. In the words of a retired professor of philosophy: "At the core of the problem is a combination of both disciplinary hubris and academic territoriality, both militating against cooperative ventures across disciplinary and administrative boundaries."[8]

Nevertheless, such ventures must be attempted. And if it is not possible to *blend* traditional courses, to compose a new interdisciplinary chorale, then let us work more creatively with the courses that are already available. I'm reminded of the impressionist artists who showed that paints needn't be combined on the palette or on the brush, but that dabs of color could be so juxtaposed and intermixed on the canvas that they would blend "in the air," as it were, blend in the eye, in the imagination. As an engineer, I'm particularly reminded of Georges Seurat, the so-called "pointillist" who painted using small circles of pure color, each about the size of a lead-pencil eraser. Up close, his paintings are a conglomeration of colorful dots. Step back and you see the beauty of the scene. Just so might engineers create splendid curricula out of diverse courses, with each institution working in its own imaginative style.

Stephen Sondheim's musical play, *Sunday in the Park with George,* is about Georges Seurat, and in a larger sense about art and creativity. In the wonderful scene that concludes the first act, we see George creating his majestic painting, "Sunday Afternoon on the Island of La Grand Jatte." As the scene begins, the people on the island are milling about, arguing among themselves in a most unpleasant way. Suddenly George calls out, "Order!" And then, as the people become quiet, and magically move into graceful poses according to the artist's bidding, he calls out again: "Design." "Tension." "Balance." "Harmony." The beautifully conceived picture takes form in front of the audience's eyes, the music swells, and the curtain falls.

I fantasize that in each of the nation's engineering schools, administrators and faculty will do something like this—look out at the fragmented, ill-humored, ungracious state of our departmentalized campuses and command, "Order," "Design," "Tension," "Balance,"—and finally—"Harmony."

## EATING CAKE

Making engineering education more congenial, mainly by eliminating the boot camp approach, humanizing the curriculum, and enhancing faculty incentives, will be a big improvement. But how can we also accommodate the constantly increasing amount of technical material with which engineers, ideally, should be acquainted? As technical discoveries proliferate, and as the number of engineering disciplines expands—computer, biomedical, environmental, and so forth—the amount of "essential" engineering knowledge grows to unmanageable proportions. This is sometimes characterized as the problem of stuffing ten pounds of material into a five-pound bag.

The traditional response has been to call for a longer program, to increase the course for the entry-level engineering degree from four years to five, or even six or seven. In addition to giving engineering students more time to assimilate the ever-expanding fields of knowledge, this would elevate the status of engineering by bringing it more in line with those professions, such as law, that feature graduate school education.

But as most engineering students and their families already find it difficult to pay for a four-year degree, the idea of requiring them to pay for more—while delaying their earning career as well—is somewhat reminiscent of Marie Antoinette. (That French queen, it will be remembered, upon being told that the peasants had no bread, is said to have replied, "Let them eat cake.")

I have visited too many engineering schools, and spoken to too many engineers and engineering educators, to believe that this worthy concept—lengthening the degree program—can be put into universal practice anytime soon. I wish it were otherwise. But an adequate constituency for such a momentous change simply does not exist. What *can* be done, however, is for each institution to reassess its current programs, to craft new ones featuring a basic four-year curriculum as rich and inspiring as it can be, and to encourage students to pursue at least a one-year master's degree. About a third of recent engineering graduates continue their studies past the bachelor's level. As this percentage increases—which each educator and employer should zealously advocate—the idea of a longer basic course will begin to take hold.

Industry must play its part by starting to compensate engineers appropriately for the additional investment of time and the additional talents developed in postgraduate work. Unhappily, there is no assurance that such an enlightened policy is in the offing. Engineers may have to be more assertive, and employers will have to become more farsighted. But the engineers may have to make the first move by proving themselves, which they can do best, of course, by *im*proving themselves.

Select schools have already started experimenting with longer programs. Their experiences will provide valuable guidance as the engineering profession strengthens and matures. While the worthy masses are doing their best to thrive on bread, a few of their number will be reaching—not for cake—but for the nourishment of an enriched professional education.

CHAPTER 12

# TOWARD THE NEW MILLENNIUM

## WITNESS TO AN AGE

When I was a boy, I usually spent the hour between 5:00 and 6:00 P.M. doing my homework while listening to four quarter-hour radio programs: "Little Orphan Annie," "Renfrew of the Mounted," "Jack Armstrong (The All-American Boy)," and "Buck Rogers." My favorite was "Buck Rogers," the *Star Trek* of its day. (This may not seem consistent with my criticism of *Star Trek* in Chapter 2, but then I was a boy.) I recall that the program started with a booming declaration: "Buck Rogers—in the twenty-first century!" Oh, how I wanted to live to the year 2000 when rocket ships and ray guns would be the stuff of everyday experience. And how unthinkably distant was that special-sounding year.

Somehow the time has passed, and the new millennium approaches. The actuarial tables (which have become increasingly encouraging, thanks to technology) say that I should have my wish. But whether or not I reach that magical date and beyond, I feel blessed in many ways, not least for having been an engineer during the specific span of years allotted to me.

The historian Arnold Toynbee said that if he had been given his choice of societies in which to live, as a citizen and family man, he would have chosen the Dutch Republic at the height of its glory in the seventeenth century. As a historian, on the other hand, he would have elected to travel with Alexander the Great. I have not given much thought to when I would have chosen to live as a citizen and family man except to note that it has been comforting to be a father after the discovery of antibiotics. As a civil engineer, it might have been nice to build aqueducts in ancient Rome or railroads in America during the 1860s. But as an engineer in general, and particularly as an observer of the engineering scene, I could not have wished for better fortune.

I grew up during the Great Depression, a time when the salvation of humankind seemed to depend upon such majestic works as the dams of the TVA, and when hopes for the future were embodied in the General Motors Futurama exhibit of the 1939 New York World's Fair. It was a wonderful time to dream of becoming an engineer. Toward the end of World War II, I served in the Seabees, not as a combatant but nevertheless proud to be in the company of those men who coined the motto, "Can Do!" Shortly after the war, I worked for a while in Venezuela, helping build facilities for the mighty Esso Oil Company. My contemporaries and I embarked on our engineering careers at a moment in history when the United States was incredibly strong and rich, when technology bid fair to provide the answer to all problems, and when the prospect of nuclear power evoked talk of energy that would be "too cheap to meter."

Of course, things did not work out the way we thought they would. Postwar euphoria gave way to "the age of anxiety," and the engineering profession felt in full measure the change of mood. In spite of the fabulous achievements of engineers—space exploration, computers, fiber optics, and much more—members of the profession found themselves blamed for the arms race of the 1950s, vilified by the counterculture of the 1960s, and called to account during the environmental crisis of the 1970s. Also, the world dominance of the United States began to erode. Engineers of my generation, along with the rest of society, were forced to reevaluate many of their preconceptions about the nature of progress and about the destiny of the United States.

Happily, in the 1980s and 90s we have seen the evolution of a more balanced attitude toward technology, consisting of both hope and caution, and tempering ambition with aesthetic and moral concern. Also, we have begun to see how closely our fortunes are intertwined with those of other lands, and how our technological efforts are affected by international competition and cooperation. Having run the gamut from the naive enthusiasm of our youth to the chastened determination of our maturity, we find ourselves still active at a time when engineering has regained some of its former public esteem, indeed when engineering may very well be on the brink of a grand renaissance.

A little younger and we would not have been engineers during and immediately following World War II, a time that, for all its ingenuousness, was thrilling to experience. A little older and we would have ended our careers while our profession was being widely denounced and our national confidence at low ebb, and before we could bring to bear the lessons we learned during those years of travail.

I do not claim that we are wiser than our fellow engineers, junior or senior, or than other individuals of whatever calling or age. In fact, I am apprehensive lest the "realism" we have learned through bitter experience be turned against the idealistic vision that we look for in each younger generation. (Such vision seems diminished enough in this time of political and economic conservatism.) However, we *should* be wiser as a result of what we have experienced, and we owe it to ourselves and to our fellows to share whatever insights are the residue of our journey through time.

No mature adult truly enjoys recognizing that he is older than he used to be. But all things considered—and considering all things is, after all, what the engineering method is all about—I'm happy to be a witness to my own time.

## CONSTRUCTIVE COLLABORATION

One of the fascinating aspects of having lived during the period following World War II is to have seen at first hand the destruction, and then the resurrection, of Germany and Japan. The amazing recovery accomplished by these nations is a testimonial to the human spirit and especially to the importance of engineering talent. Unhappily, the recovery of our former enemies, and the relative decline of our own economy, has led to renewed confrontation.

We speak of economic and technological competition as a force for good in the world, and so it is. However, unless each nation feels that it is getting a "fair shake"—and a fair share—competition can breed hostility. Along with healthy competition, we must find ways to foster cooperation and cultivate good will.

Perhaps the most worrisome example of international economic antagonism today is between the United States and Japan. Leaders of industry and government express ill will in a number of ways, and pundits have filled the bookstores with forecasts of calamity. For American engineers in particular, Japan arouses disturbing emotions: admiration verging on envy, apprehension tinged with anger.

Yet public attitudes can be influenced by personal encounters as well as by headlines. That is the belief that underlies various "people-to-people" ventures, a belief that helps sustain hope even when prospects for harmony look bleak. I know that my own ideas about Japanese-American relations are forever affected by my experience during the months immediately following World War II.

In November 1945, I was sent with the 29th U.S. Navy Construction Battalion to Truk, an atoll in the Caroline Islands that had been bypassed by the American fleet on its westward campaign across the Pacific. A coral reef surrounding the large lagoon had made an invasion infeasible, so the major Japanese naval base was neutralized by intense assault from the air. When the war ended, forty thousand Japanese servicemen, desperately short of food and supplies, lived in facilities that had been reduced largely to rubble.

The surrender terms for the base were unique in that instead of becoming prisoners, as was typical throughout the Pacific islands, the Japanese on Truk were deemed "disarmed military personnel," permitted to keep their uniforms, maintain their own discipline, and live in their own camps. In return, it was agreed that about three thousand of their men would remain on the island for several months to help reconstruct the destroyed facilities.

So it came to pass, that I—a twenty-year-old ensign fresh out of engineering school—found myself in charge of a Japanese construction crew. While the senior officers of our battalion busied themselves with such major projects as rebuilding the airstrip, I was given two dozen or so Japanese soldiers, under the command of one of their lieutenants, and sent into the hills to build a small dam. The purpose of this structure was to impound the water of a mountain stream for pumping into a new water supply system. Three young enlisted men were assigned to go with me, and we four were as apprehensive about the erstwhile enemies in our charge as they doubtless were about us. None of the group—American or Japanese—knew anything about building dams, and this plus our mutual mistrust and inability to understand a single word of each other's language made the enterprise appear more than a little ludicrous.

Yet the construction of earth-fill dams is one of humankind's primordial skills, as is the technique of communicating through gesture and empathy. With the help of an old civil engineering handbook and a few suggestions from my fellow engineer-officers, I prepared a rudimentary plan and we set to work.

And how we did work! The structure, as I gauge it now from recollection and faded photographs, was about fifty feet long at the top and rose approximately twenty-five feet from the bottom of a small gorge. Not exactly Grand Coulee, but to those of us to whom it was "the job," it loomed handsome and impressive. The work took about ten weeks, and during that short time our band formed ties of mutual respect and affection that would have been impossible to conceive at the outset.

It began with the enlisted men, who quickly established that boisterous camaraderie that seems universally to spring up among people who labor together in factories or on construction jobs. The Japanese lieutenant, who at first was fiercely proud and militarily correct, took somewhat longer to thaw but eventually became a special friend. I knew that we had achieved a breakthrough when he encouraged me to call him "Moe," an abbreviation of a very long name that I found difficult to pronounce.

At the dedication ceremony that marked the work's completion, Moe surprised me with the gift of a small statue of a revered Japanese naval hero along with a message inked on a white kerchief. The English text was a painstaking translation prepared by one of Moe's fellow officers:

> April 2, 1946
>
> The souvenir of the water plant completion. This is a statue of Admiral of The Fleet Count Hehachiro Togo, I.J.N. He was born at Kagoshima in Kyushu about a hundred years ago. He won the great victory in the Naval Battle of the Japan Sea. Namely he defeated the great Russian fleet (the Barutic Fleet). But some years ago he had a natural death. The world people say that he is the Nelson of the east. I pray that you may be able to make a great work as well as his achievement.
>
> Lt. Moe

Today the statue stands on a shelf in my office, and the kerchief, framed, hangs on the wall beside it. When I look at them, feelings of hostility toward Japan tend to fade away.

I know that this is a sentimental tale from the past, with little immediate bearing on the problems of the moment. It is also a parable that contains timeless elements of hope. The moral is clear. When people from different cultures get to know each other, fear and suspicion are likely to dissipate. Further, when they work together on constructive projects, the experience nourishes feelings of gratification and good will. Is it simplistic to suggest that collaborative efforts in engineering might show the way toward improved Japanese-American relations?

## GREATER ENGINEERING

Good will among nations is only one important element in the quest for a stable world, only one factor in the unending search for the good life. Without personal moral standards, without regard for ethics and virtue, prospects for humanity would be bleak.

According to many reproachful critics, our technical genius has far outstripped our moral sensibility. We can go to the moon and explore the atom, so they say, but we've made precious little headway toward achieving a worthy society. I have long thought that this glum view is only a half-truth, and perhaps not even that. As I look at the progress made in areas such as racial justice, women's rights, consumer protection, environmental preservation, and safety in the workplace, I can only conclude that our moral awareness *has,* by and large, kept pace with our technical achievements. Indeed, I would argue that technological progress has brought moral progress in its wake.

As I have said earlier in this book, improved public health, longer life spans, and more widely available material comforts have permitted—in a way, evoked—higher levels of education, compassion, and democratic impulse. The more securely technology protects us from the brute demands of nature, the better the chances for benevolence to evolve and flourish. I believe this to be true despite the proliferation of frightful weapons, and despite the injustices, and even atrocities, that persist in many places. Additionally, technical creativity, in and of itself, brings fascination, challenge, and hope to oppose the voices of despair.

I don't, however, find many people articulating this message of affirmation, least of all the technologists who should be its strongest champions. Occasionally one hears superficial slogans such as "Better living through chemistry," but such statements do not do justice to what has happened. With only hucksters to sing its praises, technology comes to signify the "goods life" (to use the term coined by the late Lewis Mumford) instead of the Good Life in the largest and best sense. If the reputation of technology is allowed to decline, then the quality of that technology may, as a consequence, decline as well.

It is unfortunate that engineers have generally failed to profess the social value of their work. If technologists are perceived to be taciturn—perhaps even uncaring—then the argument on behalf of technology's humane influence is certainly compromised.

Thinking along these lines, I was heartened to read of Simon Ramo's call for a "greater engineering," in which "more of the non-technological realm" is encompassed. Speaking at the symposium commemorating the 25th Annual Meeting of the National Academy of Engineering, in October 1989, Dr. Ramo—who is cofounder of the giant high-tech firm, TRW—predicted an imminent flowering of the profession, which "will come to embrace more of the issues at the technology-society interface." He spoke of the need "to harmonize technology with society" and the obligation "to meet societal demands." Most concretely, he called upon engineering educators to acquaint their students with the "social-economic-political systems that will exert decisive influences on what they will be assigned and privileged to do."[1]

This differs strikingly from the speeches usually delivered on such occasions. The traditional engineer-orator stresses the technical achievements of the past and calls for improved technical proficiency in the future. We often hear scientists make grand pronouncements about human welfare. We are accustomed to thinking of scientists as intellectuals and philosophers. Engineers have been largely absent from the arenas of civic dialogue. Dr. Ramo's concern—as well as that of the other speakers at the symposium—struck me as timely and heartfelt. I don't want to read too much into a single occasion, but perhaps on the twenty-fifth anniversary of the founding of the Academy, a number of eminent members of the profession resolved—whether deliberately as a group, or intuitively as individuals—to enunciate a more visionary message than has been customary, and by so doing, affirm their rightful place as concerned leaders of society. As noted in the Preface, this theme is beginning to resonate throughout the engineering community.

I look forward to hearing more about the link between technical progress and moral improvement, and about the evolution of "greater engineering." It is high time for engineers to step forward, assert the moral grandeur of their enterprise, and help assure the continued application of technology to worthy ends.

## AN IDEAL PROFESSION FOR IDEALISTS

As we envision an engineering renaissance and work for its advent, it is important to nurture an aspect of the profession that is incorporated in our codes of ethics but often overlooked in practice. I refer to idealism.



In the era of the Yuppie—and especially in the era of the unemployed Yuppie—it is good to be reminded that youthful idealism still burns, albeit less brightly than a few decades ago. The goal of "helping others in difficulty" is deemed important by a smaller percentage of today's college freshmen than in the 1960s, but the total number of altruistically motivated youngsters is still reassuringly high.[2]

I am disturbed, however, to encounter so many idealistic students who think they can best follow their star through a career of oversight and policing. Their role model is Ralph Nader—or else a prosecuting attorney, an investigative reporter, or a crusading politician. They seem to regard the control of evil as more useful and satisfying than the creation of good.

Even where the objective is to help the needy rather than punish the guilty, their dream is usually to become a social worker or volunteer for Legal Aid rather than, for example, an engineer who might help the disadvantaged in a material way. As I have noted previously, too many of our best young people are studying law or attending schools of social work, ignoring the fact that physical need still lies at the heart of most human misery. Engineering schools lose by default many largehearted youngsters who, if they fortified their virtue with technical training, could be a tremendous force for good in the world.

I am deeply moved whenever I think about Benjamin Linder, the young American engineer who was killed in 1987 during a battle in the Nicaraguan civil war. Linder received a mechanical engineering degree from the University of Washington in 1983. According to one of his professors, he entered the engineering program specifically to acquire skills that would help him to help others. "His whole reason for going to engineering school," said a friend, "was to go to Nicaragua so he could improve people's quality of life."[3] He designed and helped build small hydroelectric plants, and the *Time* correspondents who investigated the much-debated circumstances of his service and his death reported that "Because of his efforts, the hamlet of El Cua now has electricity."[4]

I devoted one of my *Technology Review* columns to this incident, summarizing my feelings this way: "The drama and controversy surrounding the death of this one young man in a remote jungle will have served a worthy purpose if it calls attention to the need and opportunity to blend idealism with engineering."[5]

I sometimes get the feeling (I suppose all writers do) that one of my essays has dropped into the public realm like a pebble in a dark pond, sinking to the bottom without making a ripple. This, however, was not one of those instances. I heard from a number of engineers, most notably a Peace Corps volunteer in Sierra Leone who was proud to confirm that he and many of his fellow MIT graduates were anxious to "do good"; and an Environmental Protection Agency employee in Cincinnati who wrote: "Thanks for the reminder that ours is indeed a noble cause." (There were also correspondents displeased that an American had helped peasants suspected of supporting the Sandinista cause.)

About a year later, I received a letter from the Awards Committee of the IEEE Society on Social Implications of Technology. The committee had presented its Award for Outstanding Service in the Public Interest posthumously to Benjamin Linder, and wanted to let me know that my column had influenced their decision. Considering my often skeptical feelings about champions of engineering ethics (see Chapter 9), it was somewhat ironic that I should have had any role at all in this affair. But my column was not about ethics in the platitudinous sense. The Linder story had nothing to do with rhetoric. This was an instance of real service in the real world, and I was gratified to be associated with it even in a tangential way.

There are, to be sure, many engineers who contribute their talents, unheralded, to constructive work in godforsaken parts of the earth. U.S. military engineers, serving in Bosnia with the United Nations High Commissioner for Refugees (at the request of the U.S. State Department), find other engineers serving with the World Health Organization, the International Rescue Committee, and the foreign relief branches of numerous national governments such as the U.S. Office of Foreign Disaster Assistance. The French organizations, *Médecins Sans Frontières* and *Médecins du Monde,* renowned for their prompt and daring response to emergencies in all parts of the globe, contain within their ranks not only medical personnel but also engineers to cope with the primary needs of sanitation and potable water.

The Red R (Register of Engineers for Disaster Relief), founded in Great Britain in 1980, maintains a register of engineers willing to serve from three to six months in areas of urgent need. Its recruits have repaired cyclone damage in Swaziland, helped Vietnamese boat people build camps in Malaysia, and assisted drought victims in Ethiopia. Other assignments include disaster relief in Uganda, Kampuchea, and Sudan. Some of the participants are retired engineers who receive nominal salaries; others are subsidized by their employers or a variety of relief agencies. A comparable register of engineers was established in Canada in 1982, and efforts have been made to establish one in the United States.[6]

The International Executive Service Corps, headquartered in Stamford, Connecticut, maintains a "skills bank" of thirteen thousand volunteers, mostly retired executives, many of them engineers, who stand ready to assist developing nations in a wide variety of industrial enterprises. This organization, with the support of the U.S. Agency for International Development, has sponsored over sixteen thousand projects in more than 120 countries.

The Peace Corps, now in its fourth decade, increasingly looks to engineers as useful emissaries to Third World nations. Engineering Ministries International, headquartered in Colorado Springs, sends professional design volunteers to Christian missions abroad.

I often read of pro bono enterprise closer to home: Engineering students at Northern Arizona University bringing solar-generated electricity to a Navajo Reservation near their campus; Purdue students rebuilding sidewalks in the low-income areas of Lafayette, Indiana; undergraduates at Case Western Reserve fabricating devices to aid handicapped people in Cleveland hospitals; and so forth.

One cannot help wishing that even more engineers were inspired by the ideal of humanitarian public service. Equally important, idealistic young people should come to see that many of their most fervent visions of justice can best be furthered through technological assistance to the disadvantaged.

The sufferings of humanity cannot be alleviated solely, or even mainly, through politics and litigation. Wherever we look in the world we see material well-being as the essential precondition for democracy and justice. The great challenge for this generation's youth is to direct technical ingenuity to humane purposes. Toward this end, idealists will want to understand the rudiments of technology, and some of them, at least, will want to study enginering.

The problems—both technical and social—are so daunting that many young idealists lose heart. The answer to such discouragement is to make our short-term goals realistic. Even slight improvements can be supremely meaningful. The medieval sage, Maimonides, wrote that we should consider each individual—and by extension the entire world—as exactly balanced between good and evil. Thus a single act can make all the difference:

> Every human being should regard himself as if he were exactly balanced between innocence and guilt. Simultaneously he should regard the world as being in the same case. It follows then that if he performs one good deed, he has weighted the scales in favor of both himself and of the whole world, and thus brought about salvation both for himself and for all the inhabitants of the world.[7]

The world suspended in balance is a delightful engineering metaphor, with welcome implications of hope.

## DESTINY'S BAIT

The deeds of volunteer engineers—at disaster scenes or in refugee camps—are often thrilling to contemplate. Yet, not all good works are in dramatic circumstances, nor are they always based on altruistic impulse. Often good works result from the ordinary activities of people doing their daily jobs, and even, paradoxically, people seeking profit. This was brought home to me during a ceremony I attended recently.

As co-owner of a construction company, I have, through the years, appeared at many ground breakings, ribbon cuttings, and other assorted dedications. Politicians thrive on such events, and I have rubbed shoulders with many of them—including mayors, governors, congressmen, and even one president (Lyndon Johnson at the opening of the Federal Pavilion at the New York City World's Fair in 1964). Yet, I must confess that there came a time when I started to avoid these affairs whenever courtesy allowed. I found that there was a certain sameness to the proceedings, and inevitably the excitement began to wane. Building is ever-inspiring, but the ceremony attending building can be a bore. I am a contractor, and I build for a living. Like any other business enterprise, our company pursues profit. So what care I for tedious speeches that all begin to sound alike? At least that is the way my thoughts had been running for a while.

But then came this day when something happened that changed my mind and, more importantly, refreshed my spirit. I was obliged to attend the ribbon cutting ceremony for a housing project for indigent senior citizens, built by our company for a nonprofit, do-good community group in North Brooklyn. I say obliged because the organization's executive director, an effervescent lady of considerable persuasive powers, had met my wife at an industry dinner (another ritual I usually try to avoid) and extracted a promise that we would attend.

So there we were one rainy June afternoon, crowded into the building's community room, along with the project residents and other neighborhood people, listening to speeches. An abundance of speeches! There were remarks by representatives from each of the government agencies involved in the projects's funding and processing: the U.S. Department of Housing and Urban Development, the New York City Department of Housing Preservation and Development, and the New York State Division of Housing and Community Renewal, followed by the commissioner of the City Department of the Aging, with greetings from the mayor, and envoys from the local congresswoman, state senator, state assemblyman, and president of the borough of Brooklyn.

The local city councilman spoke with passion about the grassroots activism that made projects such as this one possible. The building was being named in memory of a woman who for years had worked for the good of the community, and whose dream it had been to provide safe, attractive housing for needy elder citizens. Her family was present to accept a commemorative plaque. I had never met the lady, nor her family, nor most of the people in the room; but I was touched by the outpouring of good will.

Surprisingly, the long program, instead of having a soporific effect, began to take on an inspirational quality. How marvelous, I thought, that so many people and so many institutions had worked together for such a worthy cause. Engineering was merely one element in the splendid human venture being celebrated; but it was a key element, without which the most ardent yearnings could not be realized.

Just as I was thinking along these lines, I heard my name called. We contractors were being thanked for our good work and cooperation. I looked at our project manager and construction superintendent, tough and talented individuals who apparently had melted with kindness whenever their help was sought by the pro bono sponsors or the elderly tenants. Most engineers work for profit; we also work to employ our professional talents. But that is far from the whole story. When engineering is applied to noble ends the process is miraculously enhanced.

I found myself thinking of a novel, *Roll Back the Sea,* by the Dutch author, A. Den Doolard. It deals with the rebuilding of the dikes in Holland after their destruction during World War II. The story starts with the urgent needs of people, and progresses through the work of politicians and design engineers. Finally, contractors appear on the scene, absorbed in calculating anticipated profits. But, while they think they are working to earn money, they do not understand what is really happening. Their competence and energy are needed for this great undertaking. As for profit, writes Doolard, "profit is merely the bait that destiny has offered to these calculators."[8] Their true reward is to be found in the accomplishment of life-enhancing work.

*The bait that destiny has offered.*

What was true in Holland a half century ago is equally true in Brooklyn today, and has been true always and everywhere. I think I will resume going to ribbon cutting ceremonies.

## SAIL ON

One of the notable anniversaries celebrated during the final years of the twentieth century was the quincentenary of Columbus's first voyage to what we now call America. When plans for the occasion were first announced, the prospect was for grand festivities on both sides of the Atlantic. But as the date drew closer, a fierce revisionist crusade began to develop.

The National Council of Churches, for example, characterized the "discovery" as "an invasion and colonization, with legalized occupation, genocide, economic exploitation and a deep level of institutional racism and moral decadence."[9] An organization called The Columbus in Context Coalition proclaimed the pending celebration the best political opportunity for the left "since the Vietnam War." Kirkpatrick Sale, in a widely cited book, *The Conquest of Paradise,* depicted the Columbian voyages in terms of environmental degradation, societal exploitation, and personal greed. Gary Wills summed up the mood of the contrarians: "If any historical figure can appropriately be loaded up with all the heresies of our time—Eurocentrism, phallocentrism, imperialism, elitism and all-bad-things-generally-ism—Columbus is the man."[10]

The engineering societies took little notice of the event, and at the time I thought this was most peculiar. How could engineers remain silent, knowing that technology lies at the very heart of the Columbus adventure?

All the vices and bad impulses identified with Columbus by his critics—as well as the heroic qualities hailed by his admirers—existed long before 1492. Conquest and exploitation did not begin with the Admiral of the Ocean Sea any more than did valiant exploration. What *was* new in the late-fifteenth century was the complex of technological developments that made it possible for mariners to sail directly out into the open Atlantic with reasonable hope of finding a foreign shore and returning home in safety.

A great achievement of the fifteenth century was the rapid evolution—indeed, the almost sudden appearance, according to naval historians—of the three-masted, lateen-rigged caravel, a sturdy, oceangoing vessel that could make acceptable headway against the wind as well as with it. There were notable advances in the science of navigation, and skilled artisans produced vastly improved maps, charts, compasses, and instruments for celestial observation. In various port cities, record keepers accumulated empirical knowledge about winds and ocean currents. A veritable center of R&D in ocean sailing flourished at Sagres in southwest Portugal, sponsored by Prince Henry the Navigator. After Portuguese explorers made their way gradually down the coast of Africa, rounding the Cape of Good Hope in 1488, the equipment and experience were in hand to justify Columbus's expedition to the west.

Of course, technology alone cannot explain the European voyages of exploration. (The Chinese were technologically advanced yet remained serenely content to let other peoples go exploring.) Also, there were geopolitical forces at work, such as the blocking of eastern trade routes by the growing Ottoman Empire. Nevertheless, until science and technology reached a particular stage of development, Columbus's venture would have been unthinkable.

In reflecting upon the meaning of 1492, I've been struck by how the yearning to explore and the impulse to do engineering appear to be related, not only in the practical world of cause and effect, but also in the realm of the human spirit. Joaquin Miller's once-popular poem, "Columbus," begins with an evocation of the explorer facing open and unknown expanses:

*Behind him lay the gray Azores.*
*Behind the Gates of Hercules;*
*Before him not the ghost of shores,*
*Before him only shoreless seas.*

How remarkably this verse resembles Robert Louis Stevenson's description, in *A Family of Engineers,* of his grandfather's feeling about the profession of engineering: "The seas into which his labours carried the new engineer were still scarce charted, the coast still dark."[11]

The poem sounds somewhat naive today, as does Stevenson's florid prose. But the metaphor is as valid—and as stirring—as it ever was. At least I find it stirring. Some people, contemplating the meaning of 1492, seem to feel differently.

Columbus's critics remind us that deplorable things have been done in the name of progress, and we can surely benefit from such admonitions. But we should also recognize that the Native Americans who greeted Columbus five hundred years ago were not residents of some mythical Garden of Eden as they are depicted by a few revisionist historians. They had their own traditions of adventure, exploration, and technical ingenuity, being descended from peoples who journeyed across a land bridge from Asia and settled throughout a vast continent. Anyone who has read about the peoples Columbus called Indians, from the hunters and warriors of the northern forests and plains to the builders of the fabulous empires of the central and southern continent, knows that they were stirred by yearnings in many ways comparable to our own.

Looking back over recorded history, and even into the mists of earlier times, we perceive at the heart of human nature a passion for discovery and invention. We can explain this passion, if we like, in terms of evolutionary science—noting the survival value of creativity. We can revel in it, as does many an engineer. We can deplore it, calling it, in the dramatist's terms, our tragic fate. We cannot, however, suppress or repudiate the essence of our humanity. The lion hunts, the sheep browses, and the hummingbird sips nectar. Human beings are explorers and engineers. We are many other things as well: dreamers, artists, lovers, athletes, jokesters, and philosophers. But it is our destiny ever to embark, like Stevenson's engineer grandfather, on seas "still scarce charted, the coast still dark."

Engineers can profitably heed critics who express reservations about progress. But in our heart of hearts we echo the pulse-quickening refrain with which Joaquin Miller's lyrically conceived Columbus answers his apprehensive crew:

*Sail on! Sail on! Sail on and on!*

# CODA

In early 1975, at a time when I was deeply absorbed in writing *The Existential Pleasures of Engineering,* my wife presented me with a gift. It was the original of a cartoon by B. Tobey that had appeared in the *New Yorker* magazine. The drawing shows a middle-aged man in pajamas and bathrobe, standing in front of his house, gazing wistfully at the night sky, while his wife addresses him anxiously from an upstairs window. "All right," she says, "but promise you'll come in and go to bed the instant you *do* discover the meaning of it all."

Twenty years later, as I reach the final pages of this extended essay, I suppose I'm still looking for meanings that remain elusive. Yet there comes a time to ease up on philosophizing, to celebrate the simple pleasures, to come in at day's end and go to bed. It is also important to get up in the morning and go out to do the work that needs doing.

Living, working, and *thinking and talking* about living and working—ideally we should live in a holistic continuum uniting all aspects of human existence. Practically, this isn't possible, especially as we move ever farther from "primitive" civilization, which as I have said, appears to be our destiny. So we must strive to balance our interests and activities, to avoid shortchanging ourselves or the world by neglecting any part of what makes for the good life.

I hope that engineers, often blessed with engrossing work, and caught up in the tumult of everyday affairs, will cultivate the introspective mode. For non-engineers, I hope that there will be increased understanding, respect, and even admiration for technology.

Through such a subtle shift in our world view, we can help realize the goals I described in the Introduction and have championed throughout the book: We need more of our best young people to enter engineering, to bring the profession to new heights, and to serve society's urgent needs. We need for our political leaders, and for the public in general, to provide the support without which engineering cannot realize its potential. Our culture will be enriched to the extent that it embraces engineering and relates to it constructively. (This is also our best way to "control" technology and minimize its harmful effects.) Finally, our communal debates will benefit from reliance upon the engineering approach—logic and good sense in place of irrational invective.

Lewis Mumford once wrote: "We have yet sufficiently to realize that the symphony orchestra is a triumph of engineering."[1] I would go farther and say that we have yet sufficiently to realize that much of what is good and beautiful in our lives—to say nothing of what is indispensable—is a triumph of engineering. Increasingly, the quality of our lives will reflect the quality of our engineering.

# ACKNOWLEDGMENTS

When it comes to saying thanks, I have to "round up the usual suspects"—which means my editors, Ginny, and my family.

This is the fourth book I've done with Tom Dunne at St. Martin's Press, and for twenty years, as I've watched him move up in the firm (richly deserved) and add his dachshund logo for Thomas Dunne Books (deftly conceived), it's been a wonderful and warm relationship.

I'm grateful also to Steve Marcus, editor of *Technology Review,* and to Herb Brody who edits my column there. It's a pleasure working with such fine people and being associated with such a deservedly respected publication.

Virginia Crowley, for more than thirty years, has been helping me in dozens of ways while at the same time running the office at Kreisler Borg Florman Construction Company. Ginny complains that she doesn't get to type my manuscripts anymore, as she did before the word processor era; but she still organizes, handles notes and references, and in general makes the world a better place.

My wife, Judy, thoughtfully affords me precious time to write. Most important, she is a beloved partner—and an astute critic.

My New York son, David, applies his lawyerly skills to finding me materials of interest and trying to dissuade me from extreme and untenable positions. He also did a splendid job of editing this manuscript. Any remaining imprecision, garrulousness, or hackneyed idiom represents a victory of my stubbornness over his discretion.

My Boston son, Jonathan, is a doctor whose environmentalist convictions (along with those of my daughter-in-law, Lissa) are moral as well as medical—making me think ever more earnestly about the meaning of "progress." My granddaughters, Hannah and Lucy—to whom I have dedicated this book—also help focus my thoughts about the future.

I fear that I have said more by way of acknowledgment than the reader cares to know. Yet one doesn't contemplate "technology" in the abstract. This is a topic in which family considerations matter a lot; so I hope that even the technical-minded reader will indulge these few personal reflections.

One final word. During the four years I worked on this manuscript, I served on the Board on Engineering Education, established by the National Research Council in 1991. That service reduced the time available for other research and writing, and inevitably delayed the completion of this book. But I wouldn't have had it any other way. The Board's meetings with a wide variety of engineering groups—and the follow-up symposia in various parts of the country—were a unique opportunity to meet interesting people and to examine the state of American engineering. The Board members were most amiable companions, and I thank them (with special thanks to Chairman Karl Pister and Acting Director Kerstin Pollack) for a memorable experience.

# NOTES

## PREFACE

1. *A Profile of the Engineer: A Comprehensive Study of Research Relating to the Engineer* (Research Department of Deutsch & Shea, Inc., 1957).

## INTRODUCTION

1. Gallup Poll, commissioned by the American Consulting Engineers Council (ACEC). Press release from ACEC (20 February 1990).
2. Howard Wolff, "How Engineers View Themselves," *IEEE Spectrum* (April 1993), p. 25.
3. Cited in T. R. Reid, *The Chip* (New York: Simon & Schuster, 1984), p. 196.

## CHAPTER 1

1. Samuel C. Florman, "Sound Strategy for Competitive Cooperation," *The Scientist* (30 November 1987), p. 20.
2. Joseph Bordogna and Paul Herer, "Integrative and Virtual Partnerships" (Washington, D.C.: Directorate for Engineering, National Science Foundation, 1994), p. 1.
3. Edward F. Denison, "Accounting for United States Economic Growth, 1929–1969" (Washington, D.C.: Brookings Institute, 1974). Also, Linda R. Cohen and Roger G. Noll, with Jeffrey S. Banks, Susan A. Edelman, and William Pegram, "The Technology Pork Barrel" (Washington, D.C.: Brookings Institute, 1991). Both references cited by Bordogna and Herer, see note 2 above.

4. *Budget of the United States Government, Fiscal Year 1996; Analytical Perspectives, Budget of the United States Government, Fiscal Year 1996; Historical Tables, Budget of the United States Government, Fiscal Year 1996* (Washington, D.C.: U.S. Government Printing Office, 1995). Also, Intersociety Working Group, *AAAS Report XX: Research & Development, FY 1996* (Washington, D.C.: American Association for the Advancement of Science, 1995). There is some inconsistency between figures in these documents, mostly attributable to including capital expenditures for R&D in some tables and excluding them from others. Also, the figures used represent budget authorization rather than funds actually expended. However, for the purpose of this work, the figures are more than adequate.
5. "R&D Scoreboard," *Business Week* (27 June 1994). In this reference, aerospace and defense are noted to have accounted for slightly more than 5 percent of industry spending for R&D. However, I am assuming that small portions of R&D in other industries, e.g., automotive, chemicals, electronics, computers, and telecommunications, are also defense-oriented.
6. "Crippling American Science," Editorial, *New York Times* (23 May 1995).
7. *Science* (9 June 1995), p. 1428.
8. Barbara A. Mikulski, "In the National Interest," *ASEE PRISM* (May 1994), p. 22.
9. Robert M. White, "In Search of a Technology Strategy," *The Bridge,* National Academy of Engineering (Winter 1992), p. 7.
10. "The American Economy, Back on Top," *New York Times* (27 February 1994).
11. National Critical Technologies Review Group, *National Critical Technologies Report* (Washington, D.C.: Office of Science and Technology Policy, March 1995), p. 161.
12. Alvin Toffler and Heidi Toffler, *Creating a New Civilization* (Atlanta: Turner Publishing, 1995).
13. James Atlas, "What is Fukuyama Saying?" *New York Times Magazine* (22 October 1989), p. 38.
14. "Free to Speak, Bulgarians Aren't Sure What to Say," *New York Times* (15 July 1990).

## CHAPTER 2

1. "Profits, Reruns and the End of 'Next Generation'," *New York Times* (24 July 1994).

2. John F. Kasson, *Civilizing the Machine: Technology and Republican Values in America 1776–1900* (New York: Grossman Publishers, The Viking Press, 1976), Chapter 4, "The Aesthetics of Machinery," pp. 139–180.
3. "Machine Tools at the Philadelphia Exhibition," *Engineering* (26 May 1876), p. 427, cited by Kasson, see note 2 above.
4. "The International Exhibition of 1876," *Scientific American Supplement* (17 June 1876), p. 386, cited by Kasson, see note 2 above.
5. Richard Guy Wilson, Dianne H. Pilgrim, Dickran Tashjian, *The Machine Age in America 1918–1941* (New York: The Brooklyn Museum in association with Harry N. Abrams, Inc., 1986), p. 85.
6. Kemp Starrett cartoon, *The New Yorker,* 1932, reproduced in *The Machine Age,* see note 5 above.
7. Tom Wolfe, *From Bauhaus to Our House* (New York: Washington Square Press, Simon & Schuster, 1981), p. 7.
8. Cited by Jeffrey L. Meikle, "Plastic, Material of a Thousand Uses," in Joseph J. Corn, ed., *Imagining Tomorrow: History, Technology, and the American Frontier* (Cambridge, Mass.: The MIT Press, 1986), p. 93.
9. Rene Dubos, "Symbiosis of Earth and Humankind," Human Ecology Symposium, New York University, 24 April 1977.
10. "New Credo for Yankee Farmers: Think Small," *New York Times* (2 September 1987).
11. "Small Farms Cultivate Way of Life, and Profit," *New York Times* (23 August 1992).
12. "Ecotourism: Can it Protect the Planet?", *New York Times* (19 May 1991).
13. Edward Everett, *Orations and Speeches, III.* Quoted in Kasson, see note 2 above, p. 175.
14. S. R. Drennan, "From the Editor," *American Birds,* National Audubon Society (Fall 1992).
15. Roger Tory Peterson, "Long after Columbus," *Bird Watcher's Digest* (September/October 1991), p. 16.
16. Ibid., p. 12.
17. Paul Brodeur, "The Asbestos Industry on Trial, Part II," *The New Yorker* (17 June 1985), p. 58.
18. Paul Brodeur, "The Asbestos Industry on Trial, Part IV," *The New Yorker* (1 July 1985), p. 78.

## CHAPTER 3

1. Lewis Mumford, *Technics and Civilization* (New York: Brace & World, 1934, reissued by Harbinger Books), p. 14.

2. A. Rosalie David, *The Egyptian Kingdoms* (New York: Peter Bedrick Books, 1988), pp. 65, 66, 104, 108.
3. Richard Shelton Kirby, Sidney Withington, Arthur Burr Darling, Frederick Gridley Kilgour, *Engineering in History* (New York: McGraw-Hill Book Company, 1956), p. 32.
4. Daniel Jean Stanley and Andrew G. Warne, "Nile Delta: Recent Geological Evolution and Human Impact," *Science* (30 April 1993), p. 628.
5. "Archaeologists' Eyes Glittering Over Treasure," *New York Times* (24 July 1985).
6. Gore Vidal, *Burr* (New York: Random House, 1973), p. 196.
7. "Brazil Wants Its Dams, But at What Cost?" *New York Times* (12 March 1989).

CHAPTER 4

1. Sigmund Freud letter to Albert Einstein, cited in Arthur Ferrill, *The Origins of War* (London: Thames and Hudson Ltd., 1985), p. 14.
2. Christofe Boesch and Hedwige Boesch-Achermann, "Dim Forest, Bright Chimps," *Natural History* (September 1991), pp. 50, 55.
3. Ibid., p. 54.
4. Kirkpatrick Sale, *The Conquest of Paradise* (New York: Alfred A. Knopf, 1990), p. 318.
5. Ferrill, see note 1, p. 31.
6. Geoffrey Parker, *The Military Revolution: Military Innovation and the Rise of the West, 1500–1800* (Cambridge: Cambridge University Press, 1988), pp. 1, 2.
7. "The Mind of a Missile," *Newsweek* (18 February 1991), p. 43.
8. Daniel E. Koshland Jr., "War and Science," *Science* (1 February 1991), p. 497.
9. "In Belfast, Prosperity Eases Catholic Nationalism," *New York Times* (6 September 1994).
10. Jimmy Carter, "Global Development: Cooperation for Development Can Prevent Somalias," *Science, Technology, and Government for a Changing World* (Carnegie Commission on Science, Technology, and Government, April 1993), p. 33.
11. "Under Wide Reproach, Serbs Assert Their Culture," *New York Times* (17 November 1992).
12. Ivo Andric, *The Bridge on the Drina* (Belgrade: Prosveta, 1945; New York: Signet Books, 1960), p. 280.

13. "Panamanians Use Technology to Balk Sensor," *New York Times* (14 February 1988).
14. "For Soviet Alternative Press, Used Computer Is New Tool," *New York Times* (12 January 1988).
15. "2-Edged Sword: Asian Regimes on the Internet," *New York Times* (29 May 1995).

## CHAPTER 5

1. "Disaster Fatigue," *Newsweek* (13 May 1991).
2. "Sudan: The 'Silent Dying'," *Newsweek* (15 April 1991).
3. Norman Maclean, *Young Men & Fire* (Chicago: The University of Chicago Press, 1992), p. 258.
4. Bill Watterson, *Calvin and Hobbs,* distributed by Universal Press Syndicate, 3 May 1992.
5. *Time* (16 April 1990).
6. John Noble Wilford, "The Search for the Beginning of Time," *New York Times Magazine* (11 February 1990), p. 29.
7. *New York Review of Books* (16 May 1991).
8. Albert Camus, *The Myth of Sysiphus, and Other Essays* (New York: Vintage Books, 1959), p. 15.
9. Steven Levy, *Insanely Great* (New York: Viking, 1994), p. 286.
10. Seymour Pappert, "The Children's Machine," *Technology Review* (July 1993), p. 31.
11. Alan Cromer, *Uncommon Sense: The Heretical Nature of Science* (New York: Oxford University Press, 1993), p. 189.
12. "Feeding a Booming Population Without Destroying the Planet," *New York Times* (5 April 1994).
13. Fred Hirsch, *Social Limits to Growth* (Cambridge, Mass.: Harvard University Press, 1976), p. 20.
14. Ibid., p. 187.

## CHAPTER 6

1. Anne Carson, "A Dangerous Affair," *New York Times Book Review* (26 June 1988), p. 42.
2. David Joravsky, "Machine Dreams," *New York Review of Books* (7 December 1989), p. 11.
3. Andrei Codrescu, *American Poetry Since 1970: Up Late, Second Edition* (New York: Four Walls Eight Windows, 1989), p. xxv.

4. Bill McKibben, *The End of Nature* (New York: Random House, 1989), p. 49.
5. Wendell Berry, *What Are People For?,* excerpt in *Utne Reader* (March/April 1990), p. 52.
6. John Updike, "Deep Time and Computer Time," *New Yorker* (7 September 1987), p. 109.
7. Saul Bellow, "Machines and Storybooks," *Harpers Magazine* (August 1974), p. 59. (A paraphrase by Bellow of a quote from Russian writer V. V. Rozanov.)
8. Charles Chaplin, *My Autobiography* (New York: Simon & Schuster, 1964), p. 144.
9. Charles Riborg Mann, *A Study of Engineering Education* (New York: The Carnegie Foundation, 1918), p. 16.
10. Aldous Huxley, *Literature and Science* (New York: Harper & Row, 1963), pp. 44–45.
11. Ibid., p. 107.

CHAPTER 7

1. *Engineering Workforce Bulletin* (American Association of Engineering Societies, April 1994).
2. *Engineering Manpower Bulletin* (American Association of Engineering Societies, October 1991), Table 2, Source: National Science Foundation.
3. James L. Adams, *Flying Buttresses, Entropy, and O-Rings* (Cambridge, Mass: Harvard University Press, 1991), p. 44.
4. Eugene Fergusen, *Engineering and the Mind's Eye* (Cambridge, Mass: The MIT Press, 1992), p. 194.
5. Samuel C. Florman, *The Civilized Engineer* (New York: St. Martin's Press, 1987), p. 76.
6. Walter W. Frey, Letter to the Editor, *The Institute,* IEEE (July/August 1993).
7. James M. Watson and Peter F. Meiksins, "What Do Engineers Want? Work Values, Job Rewards, and Job Satisfaction," *Science, Technology & Human Values* (Spring 1991).
8. R. A. Ellis, "At the Crossroads: Crisis and Opportunity for American Engineers in the 1990's," a special edition of the *Engineering Workforce Bulletin* (Engineering Workforce Commission of the American Association of Engineering Societies, January 1994), p. 7.
9. "Engineers Are Finding New Directions for Skills," *ASME News* (American Society of Mechanical Engineers, March 1994).

10. Robert K. Weatherall, "The Job Market for Engineering Graduates in 1993," *Engineering Workforce Bulletin* (Engineering Workforce Commission of the American Association of Engineering Societies, December 1993).
11. Alex Taylor, III, "What's Ahead for GM's New Team?" *Fortune* (30 November 1992), p. 60.
12. "Court Room Drama Pits GM Against a Former Engineer," *New York Times* (19 January 1993).
13. Quoted in: "Harrassment Charge Poses Threat to Leadership of Drexel U. Head," *New York Times* (7 June 1987).
14. *New York Times* (26 January–30 January 1986).
15. "Superconductivity Warming Up," *The Institute* (IEEE, April 1987).
16. "Superconductor Claim Raised to 94 K," *Science* (6 March 1987), p. 1137.
17. "Two Surveys Indicate Public Holds Engineers in High Regard," *Engineering Times* (March 1994).
18. Reid, see note 3 in Introduction, p. 196.
19. Ibid., p. 196.

CHAPTER 8

1. "Can an Engineer Run Bush's Team?" *The Scientist* (26 December 1988).
2. Tom Wicker, "A Failure of Politics," *New York Times* (16 January 1981).
3. "Misleading Echoes of Carter in Clinton," *New York Times* (19 July 1992).
4. Quoted in: "Experts See '76 Victory as Carter's Big Achievement," *New York Times* (8 January 1981).
5. Paul Taylor, President of the National Council of Examiners for Engineering and Surveying, "Viewpoint," *Engineering Times* (January 1993).
6. Harold V. Rodriguez, "How Many Engineering Colleges Require Students to Take the FE Exam?" *Engineering Education* (April 1989). Perhaps the figures have changed somewhat since this survey, but I do not know of any appreciable improvement.
7. Russell Baker, "March of the Engineers," *New York Times* (13 November 1990).
8. Russell Baker, "Kindly Toward Arithmetic," *New York Times* (15 December 1990).
9. "Survey of Journalists' Perceptions of Engineers, Physicians, and Scientists," Appendix B of *Support Organizations for the Engineering Commu-*

*nity*, part of *Engineering Education and Practice in the United States* (Washington, DC: National Academy Press, 1985).

CHAPTER 9

1. Langdon Winner, "The Indifference of Technocrats," *Technology Review* (February/March 1993), p. 65.
2. Michael Walzer, "The New Masters," *New York Review of Books* (20 March 1980).
3. Interview with manager of the new product venture arm of Arthur D. Little. Samuel C. Florman, *Blaming Technology* (New York: St. Martin's Press, 1981), p. 19.
4. Edwin T. Layton Jr., "Engineering Needs a Loyal Opposition," *Business and Professional Ethics Journal* (Spring 1983), p. 57.
5. John Kolb and Steven S. Ross, *Product Safety & Liability: A Desk Reference* (New York: McGraw-Hill, 1980), reviewed and quoted by J. D. Kane in *Engineering Times* (July 1980).
6. Paul T. Durbin, "Commentary," *Business & Professional Ethics Journal* (Spring/Summer 1985), p. 147.
7. "Failures in Ethics Doomed Challenger, Says Engineer," *Engineering Times* (September 1987), p. 12.
8. Andrew Skolnick, "Stamp of Honor," *New York Times* (7 January 1989).
9. Michael Brody, "Listen to Your Whistleblower," *Fortune* (24 November 1986), p. 77.
10. "Odor from a Landfill Seeps into Flower Hill," *New York Times* (4 December 1990).
11. Letter to John Jay, 1 August 1786, *Writings of George Washington,* IX, 187 (cited in Kasson, see note 2 in Chapter 2).

CHAPTER 10

1. William E. Akin, *Technocracy and the American Dream* (University of California Press, 1978).
2. Aleksandr Solzhenitsyn, *The Gulag Archipelago I, II* (New York: Harper & Row, 1974–78), p. 197. Quoted in Ingrid H. Soudek, "The Humanist Engineer of Aleksandr Solzhenitsyn," *IEEE Technology and Society Magazine* (September 1986).
3. Sergei P. Kapitza, "Lessons of Chernobyl: The Cultural Causes of the Meltdown," *Foreign Affairs* (Summer 1993), p. 10.
4. Quoted in John Huppel Weiss, *The Making of Technological Man: The So-*

*cial Origins of French Engineering Education* (Cambridge, Mass: The MIT Press, 1982), p. 167.

5. W. E. Wickenden, *Bulletin No. 16 of the Investigation of Engineering Education: A Comparative Study of Engineering Education in the United States and in Europe* (Society for the Promotion of Engineering Education, 1929), p. 129.
6. Lawrence P. Grayson, "Rethinking Engineering Education," delivered at annual meeting of the Northeast Region of the National Society of Professional Engineers (October 1989), p. 4.
7. Cecil O. Smith Jr., "The Longest Run: Public Engineers and Planning in France," *The American Historical Review* (June 1990), p. 691.
8. Raymond H. Merritt, *Engineering in American Society, 1850–1875* (Lexington: The University Press of Kentucky, 1969), p. 118.
9. *The First Hundred Years of the Thayer School of Engineering at Dartmouth College* (Hanover, N.H.: The University Press of New England), p. 48.
10. W. E. Wickenden, *Report of the Investigation of Engineering Education, 1923–1929* (Pittsburgh: Society for the Promotion of Engineering Education, 1930), p. 129.
11. C. Wright Mills, *The Power Elite* (New York: Oxford University Press, 1956), pp. 353, 9.
12. Thorstein Veblen, *The Engineers and the Price System* (New York: Harcourt, Brace & World, 1963), pp. 139–140. Originally appeared as a series of essays in *The Dial.*
13. "For IEEE Director, Lunch with the President," *The Institute* (IEEE, July 1994).
14. Kees Gispen, *New Profession, Old Order: Engineers and German Society, 1815–1914* (New York: Cambridge University Press, 1989), p. 322.
15. Jerrier Haddad, quoted in Donald Christensen, "Is the Four-Year Baccalaureate Obsolete?" *IEEE Spectrum* (August 1984), p. 29.

CHAPTER 11

1. Jeanne Kirkpatrick, President's Lecture Series, New York Public Library (11 March 1992).
2. Richard Helms, "The Peek Under the Cloak," Lecture Series, "Spies in Fact and Fiction," New York Public Library (28 November 1989).
3. Henri Bergson, *Creative Evolution* (New York: Henry Holt & Company, 1911; Random House, Inc., 1944), p. 211.
4. Material re the various schools gathered by Robert K. Weatherall of MIT in telephone conversations with placement directors; reported in *The Im-*

*pact of Defense Spending on Nondefense Engineering Labor Markets,* a report to the National Academy of Engineering (Washington, D.C.: National Academy Press, 1986), p. 83.

5. Hugh Coble, *Texethics Newsletter* (Texas State Board of Registration for Professional Engineers, May 1992), p. 4.
6. Samuel C. Florman, *Engineering and the Liberal Arts* (New York: McGraw-Hill, 1968); *The Civilized Engineer* (New York: St. Martin's Press, 1987); "Toward Liberal Learning for Engineers," *Technology Review* (February/March 1986).
7. R.P.I. Prospectus, 1892, quoted by Ian Braley, *The Evolution of Humanistic-Social Courses for Undergraduate Engineers,* a dissertation submitted to the School of Education of Stanford University (September 1961), p. 14.
8. Steven S. Tigner, "A New Bond," *Liberal Education* (Winter 1994), p. 6.

## CHAPTER 12

1. Simon Ramo, "Engineering and Changing World Relations," *Engineering and Human Welfare: Proceedings of the 25th Annual Meeting of the National Academy of Engineering, 3–4 October 1989* (Washington, D.C., National Academy of Engineering, 1990), p. 41.
2. *The American Freshmen: National Norms for Fall 1994,* Cooperative Institutional Research Program, Graduate School of Education, University of California, Los Angeles (December 1994). 61.7 percent of college freshmen consider it essential or very important to "help others in difficulty" (p. 26); 70.1 percent "performed volunteer work" during the past year (p. 15).
3. "The Road to a Violent Death: Victim Sought 'Social Impact'," *New York Times* (30 April 1987).
4. "The Sad Saga of a *Sandalista,*" *Time* (11 May 1987).
5. Samuel C. Florman, "Engineering: An Ideal Profession for Idealists," *Technology Review* (October 1987).
6. In 1992 American Engineers for Disaster Relief (ADR) was established, a constitution and bylaws drafted, trustees appointed, and three issues of a newsletter published. Since then, according to Jim Cohen, editor of the newsletter, little progress has been made. As of late 1995, however, Mr. Cohen stated that he still had hopes that the organization would eventually grow and develop.
7. *Maimonides' Great Code, The Laws of Teshuvah,* Chapter 3, Section 4.

8. A. Den Doolard, *Roll Back the Sea* (New York: Simon & Schuster, 1948), p. 170.
9. Quoted by Charles Krauthammer, "Hail Columbus, Dead White Male," *Time* (27 May 1991).
10. Ibid.
11. Robert Louis Stevenson, *A Family of Engineers, The Works of Robert Louis Stevenson, Volume X* (New York: The Davos Press, 1906), p. 273.

CODA

1. Lewis Mumford, *Art and Technics* (New York: Columbia University Press, 1952), p. 8.

# INDEX